Fools
and
Geniuses

A novel about science (and superstition)

Surendra Verma

ISBN 978-0-646-55207-1 (paperback)
ISBN 978-0-646-55119-7 (ebook)

Cover design: Simon Kwok, Infographics

First published in paperback in 2011 by

**zscience**

ozscience@hotmail.com

About the Author

Surendra Verma is a journalist and author based in Melbourne, Australia since 1970. He has published numerous popular science books internationally, which have been translated into ten languages. His recent books include:

The Mystery of the Tunguska Fireball
Why Aren't They Here?: The Question of Life on Other Worlds
The Cause of Mosquitoes' Sorrow: Beginnings, Blunders and Breakthroughs in Science
The Little Book of Scientific Principles, Theories & Things
The Little Book of Maths Theorems, Theories & Things
The Little Book of Unscientific Propositions, Theories & Things

and a children's book
Who Killed T. Rex?: Uncover the Mystery of the Vanished Dinosaurs

Acknowledgment

This project has been assisted by the Australian Government through the Australia Council, its arts funding and advisory body.

Australian Government

iv

Historical Note

Historical records about Newton's funeral are sketchy. He died on 20 March 1727 and lay in state at the Jerusalem Chamber. The funeral was held on 28 March, and a contemporary newspaper report says that the pall was supported by six peers and the Dean read the service. The Westminster Abbey records show payments to the organist, the choir and bellringers. These records also show payment for torches, suggesting that the funeral took place at night (night funerals were popular during the early eighteenth century). Voltaire attended the funeral and was highly impressed with all the pomp and ceremony. The description of the funeral in this book is based on these and other facts but is fictional.

The last witchcraft trial in England which resulted in execution was held in Exeter in 1684. The witchcraft case in this book, set in Hertfordshire in 1683–84, is fictional but draws from historical accounts of witchcraft cases, with a poetic licence. There was a witchcraft trial in Hertfordshire in 1712 and a woman was sentenced to death but the sentence was never carried out. The last serious indictment for witchcraft was brought against a Leicester woman in 1717 but the case was thrown out of court. The old English laws against witchcraft were repealed in 1736.

The verses in this book, except one which is original, have been adapted from the literature of the period.

The facts of science and the biographical details of scientists which appear in this book have been painstakingly researched.

Fools and Geniuses

Casting the Circle

In my century, and for many centuries before, only men had the knowledge; women were either Venuses to be admired or witches to be abhorred. I really do not care what wise men say, I believe man – the gender-neutral beast of my days – spurns his natural world and yearns for superstitious, supernatural, spiritual worlds.

The tickets to the spiritual world are available from our men of the cloth, and I'm not the priestess who can open the door to the supernatural world though I'll let you enter into my weird world.

To take you there, I ought to cast a magic circle ... from the east Amaymon, king of the east; from the south Gorson, king of the south ... You're now inside the circle. No, no, I haven't cast any spells on you, but beware you cannot leave until the circle has been closed.

~~~

The euphonious splsh splt of the sculls quietened and the boat stopped at the north side of the river at Westminster Stay. 'We haven't heard the bells yet,' I said almost inaudibly, 'we're in time.' Anne gave me a quizzical glance, took out five silver

1

shillings from her purse and handed them to the waterman who mumbled thanks in gratitude for the generous fare.

'You need a new boat, sir.' She pointed at the battered boat.

'I've been plying this boat since I married me ol' woman. By your reckoning, miss, I'll need a new wife as well,' the waterman replied with a wry smile; a rotund balding man with a flowing white beard, he seemed nicer than his fellow watermen who were notorious for their quarrelsome, arrogant and greedy behaviour.

'Change both,' she said with a mischievous laugh.

The waterman shrugged and turned towards me. 'My mind's like a sieve, ma'am, but still I can ne'er forget the year of the Rye House plot to kill the king. Not that I much care about history but that was the year I married a lovely lass, bought a beautiful boat and then the river froze.'

'My mother told me it was a good fun.'

'Not for the watermen. There was no work for us. We almost starved.' He suddenly changed the subject. 'Going to the funeral, ma'am?'

'Yes.' I nodded.

'Great man he was indeed. God bless his soul.'

'So you knew of him?' I asked with a surge of curiosity sitting down again on the bench.

'Aye ma'am. People adored him. They would stand for hours for a glimpse of him in his carriage. I was lucky to see him closely when he crossed the river in my boat all those years ago.'

'Oh, really?' Anne was surprised by the waterman's chance encounter.

'The boat then had the shiny red and green paint, miss. I still remember his very lively and piercing eyes. He was of medium height, a bit plump but sharp featured; and gracious and generous like you, miss.'

Anne blushed and nudged me. 'Let's us go, Auntie Jane.'

Though it was early spring a heavy snowfall the night before had turned the river bank into a gully of grey slush. We got off the boat, held up our black wool gowns to save the hems from mud and walked towards the New Palace Yard. A coach rattled across the yard in double quick time, seeing us the coachman cried, 'Hoy.' We ignored him and entered into the narrow lane that led to the Old Palace Yard. Within minutes we were looking at the west front of the magnificent Church of St Peter. To one poet, a temple of silence and reconciliation where the enmities of a thousand years lie buried. To everyone else, the venerable Westminster Abbey.

We had little time to admire the façade; its glow slowly melted into the night as the sun closed its benediction. The elegant façade had recently been rebuilt and refaced in creamy-brown stone by the great architect Sir Christopher Wren. He had been dead for two years but his able pupil Nicholas Hawksmoor was now supervising the building of two towers which would nearly double the height of the west front. We could still see the low roof of the Jerusalem Chamber but the view would soon disappear behind the emerging south-west tower.

It's in the Jerusalem Chamber he lies in state, like a sovereign surrounded with scutcheons, sconces and silver candlesticks. In an ornate mahogany coffin, embalmed and covered in a laced flannel shirt tied at feet with a woollen thread, a linen cap on the head fastened with a broad chin cloth, woollen gloves on the hands and a cravat round the neck. His face is serene, as if he is sleeping with the gods whom he belonged. *Nec fas est propius mortali attingere Divos* ('Nearer to the gods no mortal may approach'), exactly forty years ago his friend the astronomer Edmund Halley had declared in the introduction to his magnum opus. Many of those who had read the book shared this conviction. No book had ever achieved such fame and authority, except the Holy Book.

When on Sunday morning Dr Mead, who had examined him a month before his death and had diagnosed stone in the bladder, told me about his laying in state, I left the church immediately after the service and took a hackney coach to the Abbey to pay my respects. Anne asked, 'Why?', but I didn't know how to answer her. I said that I had been attracted by a cosmic force that had momentarily gravitated to the Jerusalem Chamber. She failed to understand my silly allusion. Today is Tuesday, and I'm back to witness the historic occasion. This time with Anne who had not been to the Abbey before and was overwhelmed by the enormity and vastness of the buildings and the pomp and splendour of private coaches and their passengers which had now started streaming into the Abbey's grounds.

The six bells of the north-west tower had started tolling. They would toil mournfully every minute for eighty-five minutes, reflecting the years of the deceased's life. I held Anne's hand and started walking towards the Jerusalem Chamber. She gave me her habitual nudge. 'Auntie Jane, have we really been invited or are we gatecrashing?'

'Sssh! His half-niece Catherine Barton has personally sent me an invitation ticket.'

The Jerusalem Chamber was packed with mourners; all dressed in black, women in gowns, men in coats and breeches with silk sashes draped diagonally across their left shoulders. When I looked at the coffin which was now covered with a huge embroidered black velvet pall lined and hemmed in white silk, I remembered the words of a French mathematician who was overawed by the great book and wanted to know about its author, 'Does he eat, drink and sleep like other men?' Mortal he was, truly, but now a god in the annals of science.

At the far end of the adjoining Jericho Parlour I saw Dr Mead in deep conversation with a handsome young man who appeared to be French. They were oblivious of people around them. Servants, all dressed in black, were offering mourners cups of claret, boiled with sugar and cinnamon. As we took a cup each murmurs from the Jerusalem Chamber suggested that the funeral procession was about to move. I quickly drank the wine and whispered to Anne who had suddenly become very quiet, 'Let's go outside.' She took a sip and thrust the cup into the hands of a servant standing near the door.

Outside, the disarming darkness of the soothing spring night was leisurely letting her curtain down. Some twenty men, totally covered in black robes and hoods, were standing near the central window of the west front with heavy wax torches to accompany the funeral procession. The flames of their torches, three thick wax tapers twisted at the stem and then branching out, were flickering in the cool wind.

Eight beadles, two by two, each carrying a long staff swathed in black crepe, at the end of which was a knob of silver, led the cortège. They were followed by clergymen and other parish officers and the Dean of the Abbey. Next was the coffin, draped in the pall, carried on the shoulders of six gentlemen, three on each side. The corners and the sides of the pall were borne by six peers of the realm, all Fellows of the Royal Society, in their full regalia. Sir Michael Newton, a knight and distant relation of the deceased, was the chief mourner. He was followed by Catherine Barton and her husband John Conduitt and some other relations, other Fellows of the Society and numerous other eminent persons. Servants were handing out mourners sprigs of rosemary to be thrown into the grave on the top of the coffin. As Anne and I joined the procession, I cried in my heart thinking that such a great man had never married and had no close relations to mourn his passing.

The procession moved through the west door into the vast nave, which was filled with the sweetest music, both for the organ and for the voices, I had ever heard in a choir. Near the screen at the entrance to the choir William Croft was playing the organ and in the stalls the choir was singing the Henry Purcell anthem 'Thou knowest, Lord, the secrets of our hearts'.

The procession passed through the choir and the crossing and stopped at the entrance of Henry VII's chapel. The coffin was laid before the high alter and the choir sang the burial sentences, 'I'm the resurrection and life, said the Lord ... The

Lord gave, and the Lord hath taken away; blessed be the name of the Lord', while Croft played the music he himself had composed. The mourners stood silently. My mind had left the chapel and was engrossed in the idea that the great man was no longer bounded to the earth by the gravitational force he had discovered and now from heaven could truly appreciate how this universal force orchestrated the harmonious dance of the heavenly bodies. The universe, to this true believer, was not a blind game of chance and necessity but the creation of the Lord of all things. 'Gravity is God,' he said.

When I woke up from the reverie, I saw the coffin being lowered into a crypt in front of the choir screen and heard the Dean reading the blessing from the Book of Common Prayer. 'The Grace of our Lord Jesus Christ, and the love of God, and the fellowship of the Holy Ghost, be with us all evermore. Amen.'

The procession left the chapel, and the bells started ringing a half-muffled peal. As the custom was, the mourners assembled again. The Jerusalem Chamber now appeared as prosaic as a peacock without its feathers. The tapestry of the Henry VIII's reign adoring its cedar-wood panelled walls looked lifeless. The room had lost its spirit.

Servants were floating around carrying trays laden with cakes, pies, puddings, biscuits, cheeses, fresh and dried fruits and glasses of red and white wine. I took a slice of Banbury cake and turned around to give it to Anne, but she was nowhere to be seen. Worried I looked around and saw her talking to the young man whom I had seen earlier with Dr Mead. Anne soon noticed me and beckoned me to join them.

The young man who seemed about thirty-three bowed and kissed my hand as Anne introduced him. 'Auntie Jane, meet Monsieur Voltaire, the famous French poet.'

'Not so famous, Madame. Your lovely grand-niece is good at flattering,' he replied in perfect English.

'My French is rusty, Monsieur, but you speak so sweetly in English I won't bother trying it,' I replied.

'You would be surprised when I came in May last year I could hardly speak a word of your sweet mother tongue. Ten months on, my friend Lord Bolingbroke says, I speak it with a splendid dash and audacity.'

'French dash and audacity.'

'Of course, Madame.' He took my hand and kissed it. I admired his talent – oh, that delightful French talent – for charming women.

'Why are you in England, Monsieur?' Anne asked innocently.

'I'm an exile. Exiled from France, unjustly, absolutely unjustly. Still I'm enjoying my stay in England.'

'So you're here tonight to pay homage to Sir Isaac Newton,' I said. 'Did you know that he was the first scientist to be knighted, and he is now the first scientist to be buried at the Abbey?'

'I'm amazed to see tonight's pomp and ceremony, Madame. A funeral fit for a king. I'm incredulous that a scientist could be so honoured in England.'

'Did you meet him?' I asked.

'No. He was too sick to receive visitors. I'm so sad that I didn't get a chance to talk to the greatest creative mind our time, but I met his beautiful and charming niece and her husband,' he replied looking at Catherine Barton who was standing nearby surrounded by mourners. 'She told me many fascinating stories about him.'

Anne waved a servant to come over and took a glass of claret from the tray and gave it to Monsieur Voltaire and then pleaded in her sweetest voice, 'Could you share these stories with us, please Monsieur.'

The Frenchman seemed amused with the request but continued without any remark. 'I will tell you, Anne, how Sir Isaac discovered gravity. On a warm day Catherine Barton was having tea with her uncle under the shade of an apple tree, when he told her that the idea of gravity came to his mind under a similar situation. Upon seeing an apple falling from a tree he asked himself, Why should that apple always descend perpendicularly to the ground , Why should it not go sideways, or upwards, but constantly to the earth's centre?'

'It seems unbelievable that from such a simple observation he made the momentous discovery of gravitation.' I gushed. 'Is it really true?'

'We owe respect to the living; to the dead we owe only truth,' before he could finish the sentence he saw two gentlemen looking at him. He politely took leave of us and walked towards them.

It was also the time for us to leave. I looked for Dr Mead as we had to go home in his coach. I'm scared of walking in the dark streets of London. But when I was young I loved the dark lanes of my tiny town. Some nights Cosmo and I would wander in the lanes of Hertford looking for witches smeared with hemlock flying out of their windows on broomsticks; Cosmo clutching green and white leaves of mugwort in his hands to undo the effects of their spells. When we couldn't find any witches, with or without broomsticks, Cosmo would throw the leaves at me and dance around me singing

Well it's awful lonely for a lad to lead a single life
I think I'll go to the dance tonight and find meself a wife.

I would join him singing

Cosmo's wench turned out to be a wicked witch
Changed him into a toad and dumped into a ditch.

You didn't know Cosmo, did you? He was my physician father's young apprentice. James was his name; always talking about astronomy, my father had christened him Cosmo. Whenever my father scolded him for not doing something right, he would take a deep bow and say in his meekest voice, 'Your humble assistant is always eager to learn, Dr Digby, sir.' This always amused my father and he would humour him. 'One day you would be a better physician than me, Dr Cosmo.'

The coach was now in front of our house. We both bade goodbye to Dr Mead and went inside.

I sat on a seat in the inglenook fireplace, stoked up the logs into flames and quietly stared at them.

'You look lost, Auntie Jane, is something wrong?' Anne asked handing me a cup of coffee. 'I still don't know why you went to the funeral.'

'The funeral has stirred a storm in my mind blowing down old spider webs of sweet and sad memories. My father would sometimes tell his patients: if you share your joy with another person, you'll double your joy; if you share your sorrow with another person, you just have half the sorrow to carry.'

Anne sat next to me and took a sip from my cup. 'You dare not ignore his advice. You can share your past joys and sorrows with me, my dear Auntie.'

'Did you know I used to call him Salus? I had learned this Latin word for health and wellbeing from him when I was a child. The trickery my mother had tried to wean me away from this habit of mine would have won her many Trojan wars.'

I hesitated for a few moments, looked into her curious eyes and then continued, 'Well, like the waterman I'll never forget that year. I was as old as you. The Thames froze in December but my world froze a few months earlier ...

Raven

'Jane, wake up, Jane.'

My mother woke me up from a sweet dream ... Salus is reading a letter he has received from Oxford. His face brightens and he screams with excitement, 'Jane, the university has accepted my petition and will now admit women' ... My dream, Salus' dream!

The good doctor's ignorance of the anatomy of the female brain would have amazed swollen-headed Neanderthals calling themselves dons of Oxford, and then sipping ale from their shiny silver mugs they would have laughed at the audacity of the idea of women's mug-size brains holding as much ale of knowledge as men's barrel-size brains. Yet silly me always dreamed of becoming a Fellow of the Royal College of Physicians. Even Salus who went from mediocre Cambridge to finish his medical studies at meritorious Padua had not been a Fellow because the College won't allow more than forty Fellows at any time. I would never make the fortuitous forty. I could only aspire to become the baroness of the birthing chair: the mother sitting on the chair, many women comforting and encouraging her, midwife Jane Digby delivering the baby. A midwife could not even be trusted to deliver a baby alone; the mother's female friends, relatives and neighbours were expected – no, demanded – to be invited.

'Hurry up, Jane, it's seven o'clock, your father is waiting for you,' my mother's sonorous voice filled the house. 'Why on earth he wants to take his young daughter with him to see a patient? I can't understand it. I don't know what has gone into his head. Oh my God!'

It would be years before I could become a licensed midwife, but Salus gave me every opportunity to learn medicine.

'A girl patient for heaven's sake, Mama,' I rumbled as I got up from my bed. 'I'm not a child anymore. I'm fifteen now.'

'Watch your tongue, Miss Jane. That's no way to speak to your mother,' our old maid Martha chastised me handing me a bottle of salt. 'Clean your teeth, dress up and come to kitchen. I have hot chocolate and turkey pie waiting for you.'

'I hate cold pies.'

'It's warm, just the way you like it, Miss, er, Dr Jane.'

Martha knew how to give me fresh heart.

When I entered Salus' large but cluttered consulting room, midwife Sarah Gardiner was talking to him about the daughters of Mr Robert Bartlett, the new landlord of the Bull, an inn in Ware famous for one of its beds. The bed was so wide, said a local rhyme

Four couples might cosily lie side by side
And thus without touching each other, abide ...

Cosmo once told me that medical books were full of 'as many lies as will lie in thy sheet of paper, although the sheet were big enough for the bed of Ware'. He loved quoting from plays of some bard of Avon which he boasted about reading in a folio owned by his grandfather.

Cosmo was standing near a shelf of medicine bottles holding a bottle of bright yellow mixture and looking at it intently. I knew that he was not admiring the genie in the bottle but all his senses were centred on the conversation. I stood near him and listened to Mrs Gardiner, a well-built middle-aged, fast-talking woman, a perfect caricature of a meddling midwife and a mirror into my future.

'Last month their daughter Agnes suddenly went into a fit. Within days the fits became more regular and violent and occurred daily many times. Mrs Bartlett called me to talk about the girl's condition. When I met the girl she was as sweet as pie and a picture of perfect health. After a few minutes she started staring at me and then suddenly fell down on the ground and went into a violent fit. She was crawling and tumbling, head bobbing, hands moving violently, eyes rolling up, teeth grating and gnashing.'

'How old is the girl?' Salus asked.

'About thirteen, sir.'

'Looks like that the poor girl suffers from hysteria or epilepsy.'

'Initially I also thought so. But let me complete my story, sir.'

'Sorry, go on.'

'While the girl was in her fit all of a sudden her stockings and garters slipped down. Then I heard a little noise and her skirt and petticoat became untied and started slipping down. I can swear the girl didn't touch them. It looked as if someone was pulling her clothes down. For a few moments I was scared, sir, really scared.'

'Strange indeed.'

'The Bartletts have newly come to these parts from Cambridge. As they have much faith in their old doctor, they decided to send the girl's urine sample and a description of her physical symptoms to him.'

'A piss prophet!' Cool and calm Cosmo suddenly exclaimed uncharacteristically. 'A piss prophet!'

'I thought you were searching for aqua mentha, Cosmo,' Salus remarked.

'Sorry Dr Digby. I couldn't help but listen to this crazy story. I thought the College of Physicians frowns upon the examination of urine by physicians.'

'Yes, no good physician today will diagnose disease by examining the urine only, as the physicians did in the past. Though I believe there is place for this. We physicians are cautious by nature and slow to adopt new ways. But it's awfully rude of you, Cosmo, to call your future fellow practitioners piss prophets.'

'Cosmo is right when he calls such physicians piss prophets.' I protested. 'You yourself have told us that the ancient idea that sickness was the sign of imbalance of four bodily fluids is no longer relevant and physicians must practise experimental medicine of Renaissance physicians such as Paracelsus.'

'I have no hope of winning against these two prophets of Paracelsus. You were saying, Mrs Gardiner?'

'Yes, Dr Digby. Last week I went to see Mrs Bartlett to enquire about any news from their Cambridge doctor. As I was talking their youngest daughter Alice came in the room sneezing, screeching and groaning fearfully. There were pricks and scratches all over her face and hands. She went into a trance and lay quietly for a while. Soon after her belly began to swell and she started bouncing up her body with such violence that Mrs Bartlett and I couldn't keep her down. I swear, sir, she was stronger than any of my husband's pigs. After half an hour the girl was fine but she couldn't explain what caused scratches on her body. Agnes is ten years old, sir.'

'Strange! I have never seen such kind of seizure in my life.'

'It was the first time the girl became sick. Mrs Bartlett told me later that Alice went into a fit again when one of their neighbours had come to visit them. Mr Bartlett begs you to examine both girls.'

'Do they have any other children?'

'Three daughters only, sir. Amy is fifteen. She seems healthy.'

'Have they heard from their doctor in Cambridge?'

'Yes sir, his name is Dr William Bate. Do you know him?'

'Sure, I do. Not a cabbage-head exactly, but he certainly has utmost faith in examining urine. What was his diagnosis?'

'He wrote back saying that the girl had worms and sent a medicine. When the medicine did not help and Agnes continued to have fits, Mr Bartlett sent the second sample. This time the doctor replied that the urine sample had shown no sign of any disease and the girl was not sick of any natural cause or infirmity. He has ruled out epilepsy as there was no foaming at the mouth. He thought that her fits were probably caused by witchcraft.'

'Witchcraft!' Cosmo raised his hands and shook his head in disbelief. The bottle of mixture slipped from his hand and hit the floor splattering pieces of glass and yellow liquid all around him.

'Let's go,' Salus snapped. He took my hand and gestured Mrs Gardiner to follow us towards the coach waiting in the street. He hesitated for a moment and then said, 'Cosmo, you're also coming with us.' One of us sometimes accompanied him on his visits to nearby villages, but it was the first time he was taking three assistants with him. What was on his mind? It was definitely not the bed of Ware. He needed four more persons to check its size. Three, in fact, as we had Mrs Gardiner with us; every midwife is worth two women.

~~~

Our one-hour slow ride to Ware on the Old North Road was not pleasant. Salus was in a foul mood and did not utter a word. The rough road was crowded with wagons and packhorses carrying barley for malting. Ware was famous for its brown malt. Slow-moving wagons had churned up the boulder clay into fearsome ruts. As the horses strained slowly through the deep tracks the ride became so bumpy that at one point my father asked the coachman to stop and got down. He walked a few yards and then fell into a ditch. He cursed the road shaking his mop of tousled red hair wildly 'bloody toad-spotted measle'. Another admirer of the florid language of that bard of Avon, I presumed. I had never seen prim and proper Dr John Digby swearing like a sailor or soldier or whatever. Swearing, they tell me, not only vents frustration it also eases physical pain, and sure it did work for him. He became a bit chirpy when we turned into a country lane shaded by parallel rows of elm trees. We passed through a cluster of wattle-and-plaster huts and half-timbered thatched houses before the coach stopped in front of a large stone house with high-pitched roof. The Bartletts were very well off; apparently the Bull's big bed was not short of couples keen to share it.

We did not have to go inside the house to meet them. The whole family was under a large oak tree in the front garden, except Alice who was sitting on a high branch. She was shaking and pointing at something hidden above her in dry, brown leaves. With their eyes closed, Agnes and Amy were holding hands around the trunk and chanting, 'Don't look at her, Alice'. Mrs Bartlett was crying, 'How did you get there, darling? How did you get there?' Mr Bartlett was shouting at a maid standing near the front door of the house, 'Fetch a ladder, hurry up.'

We stood near the tree like stunned mullet and watched the unfolding drama. Alice became more hysterical and started crying, 'She is pushing me. She wants me to fall and kill myself.' Agnes and Amy started running around the trunk and their chanting became louder. Mrs Bartlett knelt on her knees and started praying, 'Our Father which art in heaven, Hallowed be thy Name.' The prayer had a strange effect on the girls. They became quiet, Alice slumped over the branch and Agnes and Amy on the ground as if struck by thunder. Mr Bartlett became frantic and ran towards two maids who appeared from the back of the house carrying a ladder. He snatched the ladder from the maids and ran back towards the tree dragging it along. He placed the ladder against the branch and brought Alice down. With her eyes closed Mrs Bartlett was now rapidly chanting in a hoarse, inharmonious voice, 'Lord Jesus Christ, Son of God, have mercy on me, a sinner.'

Salus stepped forward. 'Mrs Bartlett, I beg you to stop praying. Your daughters are under a spell and your prayers are harming them.'

Startled, Mrs Bartlett became silent, stood up and curtsied. Mr Bartlett who was still carrying Alice in his arms walked towards Salus and bowed. He was speechless. Salus stroked Alice's hair, took her hand and said, 'Don't be afraid, my child. She has gone now. Stand up.'

Alice stood up and muttered, 'Thank you, sir.'

'How did you climb the tree, my child? Did your sisters help you?'

'I flew there,' she pointed at the oak tree as a raven skirred away, 'with her, sir.'

'She can't fly back while I'm here. Be calm.'

Holding her hand he walked towards Agnes and Amy who were now sitting on the ground with their heads pressed against their chest and heads buried down into their knees. He stroked their heads one by one and said, 'Stand up girls. Let me examine you.' He monitored their pulses, looked at their tongues and said, 'There's nothing wrong with either of you. Don't be afraid of her. She can't harm you anymore.'

The girls curtsied and stood quietly.

Cosmo whispered in my ear. 'What is he talking about? God forbid Dr Digby does not believe that the girls have been bewitched.'

'He doesn't believe in such nonsense, you silly moose. He has to consider everything before he makes a diagnosis.' I made a monkey face and poked a twig I had picked up from the ground at him. 'Only half-baked doctors like James Dorrington jump to conclusions like a monkey.'

Before Cosmo could reply we heard Alice shouting and pointing at an old woman who was walking from the house opposite the street towards their house, 'Mother, look the old witch is coming to our house. Did you ever see one more like a witch than she is?' She then ran towards the women screaming, 'Take that scary cap off.'

There was a stunned silence. Mr Bartlett pleadingly looked at Mrs Gardiner. She took Alice's hand and both walked towards the house.

Horrified at her daughter's outburst, Mrs Bartlett started sobbing. She quickly composed herself and welcomed Mrs Elizabeth Marsh who lived across the road. Mrs Marsh, a slight and shrivelled woman of about sixty-four, had obviously heard Alice's shouting, but said nothing. There was nothing unusual about her dress or demeanour. The cap she was wearing, which had terrified Alice, was an ordinary knitted muffin cap.

Salus bowed and placed his hand on Mrs Marsh's shoulders. 'Ma'am, you shouldn't take what Alice said to heart. The girl suffers from some tragic affliction.' He also placed his other hand on Mrs Bartlett's shoulders and started walking towards the house.

Mr Bartlett smiled at us two bewildered bods. 'Come on in and I'll serve you the Bull's best ale. I always keep some at home for honoured guests like Dr Digby and his companions.' He seemed mighty pleased that Dr Digby was in control of the situation.

'What about serving us in the Bull's big bed?' Cosmo suggested cheekily.

'Not now, lad, but when you're married to this pretty maid,' he winked at me, 'you can have the bed for free for one night.'

True to her form, I thought, Mrs Gardiner had been offering more than medical intelligence to Mr and Mrs Bartlett.

~ ~ ~

When we arrived home after dropping off Mrs Gardiner at her house near the old castle, many patients were waiting for Salus. He and Cosmo became busy with them. Mrs Gardiner must have definitely told Mama juicy bits about witches and all that and I could see that she was dying to hear our story. But she didn't ask me; it would have meant approving my visits to patients with Salus.

I took out *The City Heiress* which I had been reading for the past few days, sat near the window and started reading it. Mama hovered around for a while and then sat down opposite me and asked, 'What are you doing, Jane?'

'Reading a book, Mama.'

'Why?'

'Reading broadens knowledge.'

'Gossiping does the same for me and it's much more fun.'

'Gossiping is backbiting. Didn't the Bible say it's a sin against charity?'

'It may be an evil pleasure, but pleasure it is. You know it makes you feel connected to others. It makes our town held together.'

'Ooo, so life is dull without titbits of gossip?'

'It's. What're you reading?'

'A play by Aphra Behn,'

'Why are you wasting your time on a play? You should be reading a medical tome by Parasol. You're always talking about him.'

'Paracelsus, Mama, Paracelsus.'

'I tell you, my little daughter, an English parasol would be more useful to you – it can save you from the sun – than a foreign doctor called Parasol.'

'I would rather carry a book by Paracelsus in my hand than a parasol. The sun would only tan my skin, but my Mama drenched with knowledge showered by the patron saint of small things would make me green with envy.'

Our banter could have gone for an eternity but it was interrupted when Salus entered the room. Mama rushed towards the consulting room saying loudly, 'Cosmo, are you still in there?'

'Yes, Mrs Digby,' came in the reply.

'Come over here, love. Martha has baked almond pudding for you.'

Mama instinctively knew that the god of gossip demands a good gathering of worshippers. She would have loved to have the whole neighbourhood around, Salus permitting.

The whole neighbourhood, in fact, the whole town would still hear the story, trickle by trickle from Mama, Mrs Gardiner, Salus, Cosmo, the coachman, the horses, anyone even remotely involved in this morning's drama. Everyone knew everyone in our town. The lack of privacy resulting in an abundance of nosiness underlined our lives. I wished Cosmo, a wizard with chemicals, could invent a scent the smell of which would scare everyone from committing a biblical sin around the wearer. I would give my right arm for a gossip-free zone around me.

Sipping ale Salus narrated the story of our visit to Ware. It took longer than the time we spent there as Mama insisted on knowing every trivial detail. She was, however, quick to arrive at the diagnosis of the girls' illness. 'I pity the poor girls. Imagine being bewitched by your neighbour.'

'I never said that they have been bewitched.' Salus quickly corrected her.

'I just put two and two together, John. I'm not talking about your medical diagnosis.'

'Shouldn't we wait for that before we start accusing someone of witchcraft,' I said angrily.

Mama ignored me and turned towards Cosmo. 'What do you think, love?'

'With great respect, Mrs Digby, to me tales of witches are no different from fairytales,' ever diplomatic Cosmo replied gently. 'Both exist only in our imagination.'

Mama smiled. 'The trouble with young people these days is that they don't read the Bible. Vicar says that the Holy Book prohibits the practice of witchcraft.'

Salus refilled his mug, took a swig and another swig and clasped his hands round the mug. 'Gentlewoman, there is no need to invoke the Bible as no one except a little girl has accused Mrs Marsh of witchcraft. While everyone was enjoying Mr Bartlett's finest ale, I was busy examining the three girls thoroughly and talking to them in private. I couldn't find any sign of physical illness, but they might suffer from some kind of mental illness we doctors do not know about.'

'I would say that their bizarre mental state has been caused by a witch,' Mama said with the confidence of a doctor's wife.

'May be you are right, Catherine,' Salus said. 'But I'm still looking at both natural and supernatural explanations.'

Cosmo's face showed a momentary look of surprise and incredulity when he heard the words 'supernatural explanations'. He was good at hiding his emotions in the presence of my parents.

'As a doctor I have been trained to be cautious,' Salus continued.

'You can be as cautious as you want to be, but it's not natural for a little girl to jump to a twenty-foot high tree branch. I'll call it supernatural. The girl said that she flew there with the witch who had turned into a raven.'

'We must at least make some effort to check the truth of the claim,' Cosmo said with a pained expression on his face. 'Why do religious people yield easily to the temptation to see what is natural as the divine, supernatural and miraculous?'

Mama piled more pudding on Cosmo's plate. 'You are not saying that you're not religious, love?'

'Sure I do believe in God, Mrs Digby, as much as you do,' Cosmo replied. 'The universe is not an accident with no purpose and no God. Life without faith would be meaningless.' He looked anxiously at me. I knew he was looking for my support.

'Believing in God does really make you feel better, sure, but believing in God is not about believing in any weirdest things,' I said smiling at Cosmo. 'Belief is a rational as well as an emotional process. For you, Mama, it's entirely an emotional process. You are one of those who are glad to be gullible. I'm not.'

Noticing that our discussion was heating up, Salus stood up. 'I have work to do.' He patted Cosmo's back affectionately. 'Take it easy, lad. We will soon find the answer to what troubles the girls, either in science or sorcery.' Cosmo winced.

~~~

A few days after our visit to Ware a harassed-looking Mrs Gardiner walked into the store room adjoining Salus' room and sat listlessly on a stool without taking off her rain-soaked riding cloak. Cosmo was mixing some stuff in a mortar with pestle to make a medicine. When he opened a bottle of arsenic to pour into the mixture, I snatched the bottle from his hand saying curtly, 'Don't you know it's a poison?' 'It's only the dosage that makes a thing poisonous, stupid,' he muttered snatching the bottle back from my hand. Mrs Gardiner didn't say a word, which was highly unusual for her not to meddle in matters medical or medicinal.

As soon as the last patient had left, she walked into the consulting room.

Salus looked surprised. 'What's going on?'

'You know Amy, sir, the oldest Bartlett daughter you saw last week? She had a seizure this afternoon. It was a terrifying sight. You won't believe a word of it if I describe it to you. It wasn't some affliction. It was the work of Satan.'

Mrs Gardiner made the sign of the cross and looked at Mama who had just walked in from the front door with a scroll in her hand. 'Mrs Digby, you were right when you said that the girls have been bewitched. I'm sure the Devil controls them now.'

'I've just been to the church and Vicar Tayler also thinks that it has to be the work of the Devil,' Mama said. Our parish church was only a few hundred yards from our Honey Lane house, handy for Mama for sharing snippets of spirituality and scandals.

Salus took a deep breath. 'I don't know.'

'Dr Digby, when could you find time to see the girl?' Mrs Gardiner asked, 'Mrs Bartlett begs you to come tomorrow.'

'No way before Monday afternoon.'

Mrs Gardiner stood up and whispered in a tired voice. 'Excuse me, sir. I've been riding whole day.'

'Come on! Sit down.' Mama insisted. 'Martha will make you a nice cup of coffee.'

'Not now, Mrs Digby. Thank you.' She walked out leaving behind a trail of water dripping from her cloak.

A shiver of silence passed through the room, instantly shattered by the swish of a scroll waved by Mama. 'It was hanging from the knob of the front door. I wonder who left it there.'

She untied the blue piece of string and unrolled the paper. She looked at it and then exclaimed loudly. 'It has funny writing on it.'

Salus took the paper from her hand and said, 'It looks like Hebrew to me.' He walked to his bookshelf, pulled out a book, looked at it. 'One of my patients gave me this Jewish Bible.' He looked at the paper again. 'No it's not Hebrew. It's not Arabic either. It could be some strange Eastern script.'

'But why would someone leave it at our door?' Mama seemed worried.

'May be it's the witch that is terrifying girls in Ware.' I teased her.

'Oh! Some patient must have left it by mistake.' Salus consoled her. 'We will soon find out.'

Cosmo took the paper from Salus' hand, looked at it for a moment, his eyes sparkled and a broad smile spread across his face. 'It's mirror writing, Dr Digby.'

'What do you mean, love?' Mama asked.

'I'll show you if you get me a mirror.'

I rushed inside and came back with my round looking glass.

He placed the paper in front of the mirror. 'Come on, Mrs Digby. Look into the mirror. Can you read it now?'

'Sure I can.' She looked pleased. 'Why would someone write like that?'

'Don't worry about that, my girl,' Salus said lovingly. He looked puzzled. 'Cosmo, please read it aloud.'

'Yes, sir.'

Weighed down by superstition? Take double dose of science. A spirit says science is a good antidote to superstition. It's all Greek to you.

Once someone asked me what was very difficult, I replied, 'To know thyself.' When she asked what was very easy, I said, 'To give advice'. Today I'll take the easy option.

Forgive me; in my eagerness to talk to you I have forgotten my manners. I'm Thales, Thales of Miletus, a philosopher.

I lived in the sixth century BC, but I'm now nothing but an illusion, just like the image you're seeing in your mirror. Why? Because my soul is restless as I find that even after two millenniums our world is far from enlightened. We have not yet learned how the natural universe can coexist with the spiritual universe. The two triangles of reason–science–reality and intuition–religion–enlightenment are still not congruent (I know something about congruent triangles for I discovered the idea of congruent triangles). These triangles are now not only incongruent but inimical to each other. Sometimes they act like arrowheads hurting each other leaving a world in which ignorance triumphs. I fear that's what is about to happen in your town. A witch's toad is already creating confusion in many minds:

A centipede was happy quite, until a toad in fun
Said, 'Pray, which leg comes after which?'
This raised his doubts to such a pitch
He fell distracted in the ditch
Not knowing how to run.

This verse reminds me of an embarrassing incident. I was also a keen astronomer and predicted an eclipse of the sun to 585 BC. One night I was gazing at the sky as I walked and fell into a ditch. A clever and pretty girl lifted me out and remarked sarcastically. 'Here's a man who wants to study stars, but cannot see what lies at his feet.' This incident gave birth to the image of an archetypical absent-minded philosopher (or professor). Yet to my contemporary Greeks I was one of their Seven Wise Men.

Absent minded I might have been, but my scientific ideas were based on observed facts not myths, though I lived in a world where mythology provided answers to almost everything. I looked for answers in nature instead of religion and gave natural rather than supernatural explanations. I would also like you to do the same.

Pythagoras, a Greek philosopher whom I met when he was a young man of twenty and whose name now appears in every school geometry textbook, came up with the remarkable idea of transmigration of soul: when we die our souls migrate to other living bodies in an endless cycle of reincarnation. He was right. My soul now inhabits a toad. Ironic it might be, but I won't let you fall into a ditch of ignorance when everyone in the town would be hysterical about some girls being bewitched. In the past such rumours have resulted in trials and execution of numerous innocent women. With right reasoning you can convince town folks of their folly.

Follow the course of reason, reason which is reasonable and respects creativity and morality. Seek answers in science, not in superstition. Let the great scientific minds be your beacon, a beacon brighter than the lighthouse of Alexandria, one of the seven wonders of the ancient world.

'Interesting isn't it?' Cosmo looked excited. 'I wonder who wrote it.'

'John did,' Mama said. 'He's pulling your legs, Cosmo.'

'You're mad to suggest that,' Salus rebuked her. 'Why would I resort to such a prank? It has to be either Cosmo or Jane.'

'It's not me, cross my heart.' Cosmo and I chorused.

'You know I can't write such cerebral crap, but I can tell you a story about this Thales. The classics mistress at my school once told this story to our class. A farmer's donkey routinely used to carry heavy bags of salt to market. One day the donkey fell into a stream, thereby dissolving much of the salt and making the burden lighter. The smart donkey learned the trick of rolling over whenever he crossed the stream. The farmer approached Thales for advice who told him to load the donkey with sponges on the next trip to the market.'

I giggled. 'I don't care who left this scroll at our door, but I look forward to receiving the next instalment from the man who can outsmart a donkey.'

Cosmo looked pleased. 'You sure there will be another scroll?'

'There won't be another scroll.' Salus snarled. 'As I said before some patient has left it by mistake. It has nothing to do with us.'

Cat

Cosmo worshipped wind-swept cleanliness. The other day when he couldn't find the bottle of sal ammoniac in the chaos of the consulting room, he lost his calm and cried. 'Dr Digby, your organised subversion of orderliness is killing me. I can't stand it anymore. Do I have your permission to create some sense in the way you shelve your medicine bottles, sir?'

Salus looked at him as if he was mad. 'No.'

'Why, sir?'

'I can find any bottle in seconds. My system may seem a bit odd to you, but it's crystal clear to me. You must learn the system.'

'Where should be sal ammoniac according to your system, sir?'

When after a long search Salus found the bottle hidden behind some bottles of ointment of St John's wort, he said sheepishly, 'You have my permission to arrange the bottles in any way you want.'

When Cosmo couldn't find time to do this job during the week, he decided to do it on Sunday afternoon. When Mama learned of his plan, she invited him over for lunch.

After a hearty lunch of roast veal, roasted wild geese, grand salad, bread, pear pie and custard followed by fruit and cheese, Salus and Cosmo adjourned to the consulting room. I tagged along.

With a glass of cognac in one hand and a tobacco pipe in the other, Salus was in a cheerful mood as he walked around the room. 'What do you reckon is the best way to arrange the bottles, Cosmo?'

'In alphabetical order, or may be according to treatment, sir,' Cosmo replied.

'What about colour or size?' I piped up.

'Why not by weight?' Cosmo said in sarcastic tone.

'Why not?' I retorted. 'Density is a good way of identifying different substances.'

When Mama who had been standing near the door with embroidery basket in her hand clicked her tongue in impatience, 'tch, tch, tch', Salus asked her, 'What is the matter?'

'Why are you all talking such nonsense?'

'Not nonsense, Catherine,' Salus said, 'simply an exercise for the mind.'

'What a dopey exercise! There is only one way to shelves bottles and that is in order of their names.'

I had to contradict my mother. 'Sure there can be other better ways.'

'John, send this darling to a boarding school, instead of you stuffing her little brain with medical nonsense.'

'Noooooo.' I screamed. 'The only thing they do is to train girls in social graces so that they can attract husbands.' I winked at Cosmo. 'I have an abundance of them.'

Salus gave a chortle. 'Social graces or prospective husbands?'

Mama curled her upper lip in disapproval. 'Neither. Anyway, what's this thing called density?'

'Sure you know, Mama,' I replied.

'I'm a clod in matters of science.'

Salus gave me an imploring look expecting me to come up with some simple meaning. 'Little pitchers have big ears, Catherine.'

'Let's see,' Mama replied.

'Density is a measure of heaviness.' I blurted.

Mama deduced as aptly as Aristotle. 'I'm heavier than you, so I'm denser than you.'

'Wiser, not denser,' Cosmo said with a chuckle of delight. He then brought two bottles from the shelf and placed them on Salus' desk. One was labelled alcohol the other mercury. 'Mrs Digby, these two bottles are of the same size and each filled to the brim. Which is heavier?'

'This one,' Mama said lifting the mercury bottle.

'Both liquids have the same volume. But mercury is heavier because it's more compact. Therefore, it has a higher density than alcohol,' he explained like a grammar school master.

Her eyebrows became denser as if she was in the drift of density up to them. 'It's still too dense for me, love. Anyway, who found out this density?'

'The French, I think,' sipping his cognac Salus said with an impish grin.

'The Frenchies? No wonder, it makes no sense to me. Frog soup muddles their mind, you know. What the world is coming to, stuffing English kids with dim discoveries made by foolish ...'

'Don't be so daft, Mama,' I interrupted. 'Density and buoyancy were discovered by Archimedes when he realised that the amount of water spilled from his bath was equal to the space occupied by his body. He was so excited by his discovery that he ran naked in the street. Everyone knows the story.'

'I happen to know a story older than the story of your mad scientist.'

I sighed. 'Not another story told by your know-it-all school mistress.'

'The story goes that a thirsty crow found a pitcher with some water in it, but he couldn't reach the water with his beak. The clever crow knew that pebbles are heavier than water, so he dropped pebbles in the pitcher until the water level rose within his reach, and he quenched his thirst. I would say that density was discovered by a crow, not your Archimedes.'

I shook my head in disbelief. 'Mmm, mmm, mmm.'

Salus puffed his pipe. 'Our Cosmo is cleverer than any crow. For peace in my house, forget about density and let the lad do his work.'

Cosmo waved his hand up and down. 'Density sinks; buoyancy soars.'

Mama pondered profoundly. 'I wonder whether witches know about density and buoyancy. They must if they can fly.'

'As much as the birds do,' Cosmo said.

'And my high-flying Mama,' I added.

~~~

Monday was the kind of day I used to call rye bread days. On this dull, damp and depressive day even my mother conspired with nature to dampen my spirits by serving cold peas porridge, stale bread and flat ale for lunch because the maid was away. Did you raise your eyebrows when I said ale? Drinking water wasn't safe in those days; young or old, everyone drank ale or beer. Didn't you know that it was the arrival of coffee on our shores from Turkey which transmogrified the sloshed English people into spirited and scheming empire builders?

When Salus asked me to accompany him to Ware to examine Amy, the oldest Bartlett daughter, I jumped with joy. Our trip on horseback – I was riding pillion – was faster and more pleasant than by coach as most of the way we bypassed the potholed Old North Road and rode through undulating fields and woods. Riding, especially for a pillion rider, is always dangerous, but the bucolic countryside, forgive the hackneyed phrase, did raise my spirits for sure.

When we arrived at their house, Alice, Agnes and Amy were playing blind man's buff on the lawn with two other girls. Alice in a blindfold was trying to catch the other girls who were running around her sometimes pushing her gently, taking care to be as quiet as possible. I felt like joining them as I had not played the game for years. When I asked Salus why the blind-folded player was called a 'blind man' when boys played the game but 'pickety witch' when girls played it, he just stared at me like the Sphinx. A red-haired Sphinx, if you could imagine one.

We ignored the girls, left the horse in the care of a servant and followed a maid to the hall. About a dozen fashionable high-backed walnut chairs were scattered around in the middle of the room and the fireplace. The lack of symmetry in furniture arrangement imposed by the need to stay close to fire made our sitting rooms look like upholsterer's or cabinetmaker's shops. One could trace the roots of many of English peculiarities and eccentricities in our fireplaces and their crackling coal-fires.

Mr and Mrs Bartlett, who had been sitting expressionless like stone effigies near the fireplace, appeared depressed, not even bothering to stand up from their chairs to welcome us. I thought how awfully rude some people could be, but Salus took a forgiving view. To this Renaissance medicine man people appeared as patients. Instead of breeding or schooling, he could ascribe bad manners to some strange sickness. Rudenitis was the name of the disease, if he didn't know. I think this godly forgiveness came to him when he took the Hippocrates oath. As I would never be allowed to take this hypocritical oath, I didn't suffer from forgivingangeletes. I was about to say, 'Sir, madam, get off your arse and greet your guests', but Salus jumped the queue and said in honey-soaked herbal voice, 'Good day, Mr Bartlett, Mrs Bartlett. Don't bother to stand up.'

He pulled up a chair between them, took out a green bottle from his leather medicine box and gave four white pills each to his new found patients with the assurance, 'They will calm your nerves.' Pills have strange effects on some people. Salus had a large bottle of sugar pills, coloured orange to disguise them. He used to prescribe them when he couldn't think of any other suitable medicine. Strangely, many patients returned after a few days, looked very happy and asked for more orangeypills; such is the power of belief on our brains. Cosmos would take out a large bottle marked 'Placere', carefully count twenty or thirty pills, place them in a paper pocket and charge the grateful patient God knows how many silver shillings. Cosmo told me that the Latin word 'placere' meant 'to please'. Pleased it indeed, both the doctor and the patient.

The pills, they were definitely not 'Placere', had the most pleasing effect on Mr Bartlett, even before he had swallowed them. He immediately stood up and left the room saying, 'I must get some ale for you, Dr Digby.' He came back, followed by a maid with four pewter tankards. 'The Bull's best ale, Dr Digby,' he made sure to remind us with a little shy smile on his face.

I don't know whether it was the pills or the ale, Mr and Mrs Bartlett were quickly cured of depression or rudenitis or both.

Empowered by the instant effects of his therapeutic skills, Salus now focused his attention on the real patient, the 'sick' girl who had been romping around outside. He held Mrs Bartlett's hand to console her. Perhaps a medical degree allowed a man to hold a lady's hand in front of her husband, an observation Cosmo later dismissed immediately by clasping his hands tightly around me, not in front of my parents though, and saying, 'You only need a degree of dare, Miss Digby.'

'Mrs Bartlett, when did you first notice any change in Amy's behaviour?' Salus asked still holding her hand.

She sobbed for a few moments and then said in a voice trembling with fear, 'Sir, Amy is a sweet girl, but when she is possessed by the demon she frightens me.'

Mr Bartlett nodded. 'Yea.'

'Amy was perfectly healthy until a couple of weeks ago,' Mrs Bartlett continued. 'One day she complained of headache, so I thought I should apply some balm. When I couldn't find any in the house I went to our neighbour across the road to borrow some,'

'Mrs Marsh, the lady I met last time?' Salus asked.

'Yes, sir. When I came back I rubbed a little balm on Amy's forehead. Within minutes she had a strange fit. For a few minutes she almost stopped breathing, her limbs became limp. Her eyes were closed and she didn't, see, speak or hear anything.'

'It could be reaction of the medicine. But it was only a balm. Did she swallow it?'

'No, sir. I ran to Mrs Marsh's house to find out whether she had given some wrong medicine by mistake. She told me it was the balm she bought from the apothecary. She then rubbed some on my forehead and it did feel like balm, a bit stinging. When I came back to the house with Mrs Marsh in tow ...' Mrs Bartlett started crying uncontrollably.

Worried-looking Salus stared at me but soon realised that his daughter might be capable of handling some minor medical emergency but was hopelessly unequipped to cope with an emotional upheaval. He then looked at Mr Bartlett, a perfect picture of an emotionless English gentleman, who sat silently and made no effort to calm down his wife. 'I wish Mrs Gardiner were here with me today.'

It dawned on me on that day that Mrs Gardiner had some skills, people skills, even my father, a doctor of medicine from Cambridge etcetera, lacked. My measure of respect for Mrs Gardiner, and by association the profession of midwifery, rose by a notch. Mind you, only by a notch.

Salus had not lost control of the situation. He commanded, 'Mr Bartlett, could you please fetch some hot coffee. Jane, come and sit next to Mrs Bartlett.'

I put my hands around Mrs Bartlett, knowing it lacked the warmth of my mother's, or even Mrs Gardiner's, hug.

Soon a maid came with four expensive sgraffito mugs. Salus took out two orangeypills and gave them to Mrs Bartlett to chew them with her coffee.

Mr Bartlett, who had returned with a small jar in his hand, gave it to Salus. 'Dr Digby, this is the balm my wife borrowed from Mrs Marsh. I had kept it as I wanted to show it to you.'

Salus looked at the jar, opened it, smelled it, and then said with a smile of relief, 'Ah, it's Dr Milton's patented balm. I prescribe it to my patients.' He then asked Mrs Bartlett who was now under the calming influence of Dr Salus' yet-to-be-patented orangeypills, 'Did you tell Amy that you were going to Mrs Marsh to borrow the balm?'

'Yes, sir.'

'When Amy came out of the fit, did she remember anything?'

'No, sir. She told me that she was sleeping.'

'When did the next attack happen?'

'When Mrs Gardiner came to our house.'

'That must have been last Friday. Have you noticed any change in her behaviour since her headache?'

'I have heard her asking 'What did you say?' a few times even when no one was around.'

'She was talking to some spirit, wasn't she, Dr Digby?' Mr Bartlett asked.

'Not really. Severe headaches can confuse the mind,' Salus replied. 'Mrs Bartlett, have you seen Amy doing anything unusual before she had the headache. Tell me even if the thing seemed trivial to you.'

'One Sunday afternoon, after we came back from the church, I saw the girls playing with an egg and a glass. Amy poured the white of the egg into the glass and was using it as a crystal ball.'

'You never told me about this, Susan,' timid, passionless Mr Bartlett suddenly roared like a lion. 'Why didn't you tell me?'

'What's so special about it, Robert?' Mrs Bartlett was equally aggressive in her reply. 'Girls sometime use it to divine their future husband's occupation. Our Amy is almost sixteen.'

'Calm down, both of you,' Salus said angrily. 'Mrs Digby, did this happen before Agnes had the first seizure?'

'Yes, sir.'

'Oh, God, our girls have tempered with the Devil's tools,' Mr Bartlett said in a loud voice. 'Did you know that divining is an occult practice, Susan? It has opened a door for Satan to enter our house and play pranks on our daughters.'

Salus seemed irritated with Mr Bartlett's outburst. 'Be quiet, sir. Let me do my job. Go and bring Amy in. I shall examine her now.'

Short, thickset Mr Bartlett could certainly bully rowdy drinkers at the Bull, but Salus' summons turned him into a sheep.

The moment Mr Bartlett left the room, Salus asked Mrs Bartlett in a very low voice, 'Did Amy see her future husband in the crystal ball?'

'No, sir.'

'Did she see anything else?'

'Yes, but I did not attach any importance to it until last week. I overheard her saying to her sisters that there is a witch in Ware and she won't let her see her future husband.' She stopped and started sobbing.

Salus placed his hand on her shoulders. 'Everything will be fine. What else did you hear? Tell me before Mr Bartlett comes in.'

She hesitated but when she saw Mr Bartlett coming in with Alice, Agnes and Amy and two of their friends, her face turned pale and she whispered. 'In her crystal ball Amy saw five girls dancing naked with a witch in the woods.'

The girls curtsied and stood in front of us silently. When I said 'Won't you introduce me to your friends, Amy?' she just raised her eyebrows in a defiant stare. Mrs Bartlett, whom I considered a well-mannered, polite and educated woman, gave a scolding look to her daughter and said, 'I'm sorry. Dr Digby, Jane, please meet Joan and Grace Pollard. They live down the road.' Joan and Grace, who were Alice's and Agnes's age, smiled cutely.

Mrs Bartlett then turned towards Salus. 'I'm deeply sorry, sir. I've not been myself lately.' Salus' medicinal smile restored her to her usual self.

'Go with Amy to her bedroom and check whether she has any skin rash,' Salus asked me. 'Better you take all the girls and check them thoroughly,'

When I came back with the girls, Mr Bartlett was sitting gloomily and silently in his chair. He had not spoken a word since he had returned with the girls. Mrs Bartlett was now radiant, chatting with Salus animatedly. Salus' orangeypills and medicinal smiles were reacting fast, overreacting, in fact. I thought she was flirting with my father. I was an overimagining, overprotecting, overlording kind of girl, you see.

'Sir, there is no trace of rash on any of them.' I announced my medical diagnosis with the confidence of a fresh medical graduate from Oxford or Cambridge.

'Good news, Jane.' Salus then pointed to a chair. 'Amy, could you please sit down here so that I can examine you.'

Salus monitored her pulse, checked her throat and ears. I could see that her face had started becoming darker and contorted. When Salus opened her right eye and came closer to look at the pupil, her face became violent and she vomited up a volley of nails on his face with such a force that he screamed, 'Eeeeee! Eeeeee! Eeeeee!'

Amy was now crawling and tumbling on the ground, grating and gnashing her teeth. After a minute she fell into a deadly trance. I was completely dumbfounded by this dramatic turn of events. Salus was in pain; his left eye bruised. He opened the medicine box took out a bottle of lotion and some bandages and handed them to me. Luckily, the nail had not hit the eyeball. The moment I finished applying the lotion and bandaging, he immediately transformed from a patient to a doctor.

Looking like a pirate with an eye patch, he sprinkled some liquid medicine on a piece of cloth and placed it on Amy's nose. Within a minute she stood up and sat on the chair as if she has just woken up from deep sleep. I was amazed at the miracle of medicine, but when I told the story to Cosmo the next day, my cynical friend said it was not the medicine but the end of Amy's role in Act 2 of the Witch of Ware. The girls were damn good actresses and deserved to perform at Duke's Theatre in London, he suggested.

I looked around the room and I counted a dozen nails. They were flat-head nails longer than a penny coin. Where did they come from in her mouth? Her face seemed quite normal when I examined her for skin rash, but she didn't speak a word.

I thought I would pick up a nail to take home to show to Cosmo. As I went around a chair to pick a nail without seeing by anyone, I saw a white cat staring at me. Seeing me approaching, the cat jumped and miaowed.

'Look Mother,' Alice quailed. 'This cat has been following us for hours.'

'This is Mrs Marsh's cat,' Agnes whizzed. 'She has been spying on us and telling her everything we do.'

Amy sneezed violently and I saw, I could swear on the Bible, a nail shooting from her nostril and hitting the cat. The cat cried, 'Meow, meow, meow', and ran out of the room.

I also wanted to run out of the room, leaving the surreal experiences of the day behind. I was glad when Salus picked up his medicine box and walked out with Mr Bartlett. Once outside, Mr Bartlett handed five gold guineas to my father and said, 'Much obliged, Dr Digby. Do you think some spirit in the children making them do so?'

'I can't find any reason for their seizures, but I shall keep on looking for answers in my medical books,' Salus replied.

'You have not answered my question, sir.'

'I advise you to see a priest.'

'We don't know anyone we can trust.'

'See Vicar Tayler. He is a highly knowledgeable pastor and a doctor of divinity. We shared lodgings when we were both at Cambridge. You will find him at All Hollows Church. Good day, sir.'

We were on our way to home, riding through the fog that had always been ready to rob our landscape of sunshine and spirits. My mind was fogged up with the disturbing thoughts about a close relationship between my father and the vicar, and ordinary people's tendency to seek solace in hocus-pocus, and the niggling notion that when I would become a midwife I would be paid not in glittering gold coins but in stained silver shillings.

~~~

In the evening, as usual, Cosmo and I met in the store room. With a playful glint in his eyes he showed me a scroll he had found hanging from the knob of the front door after we had left for Ware. Like the first scroll it was tied with a blue string.

'Did you show it to Salus?' I asked.

'Yes, but he seemed irritated,' Cosmo replied.

'Read it to me, please, I won't be irritated. You know I can't read mirror writing.'

Gazing beyond the mist of myths and trying to see the cosmos as it really is. The earth is not flat, but don't say it moves. You'll be banished.

Greetings from Thales, your guide to science, the triumph of reason over superstition.

The story of science is as vast as the cosmos, but my story is short and selective. It begins in ancient Greece with three thinkers who 'discovered' the cosmos.

Ancient Greek philosophers used reason and observation to find out about the world around them. They wanted to see things as they really are. Their methods of investigation were not flawless, their conclusions often utterly ridiculous, but their intellectual fearlessness to break boundaries of knowledge was astounding. They believed in explaining all phenomena objectively, presenting their explanations as rational arguments and distilling those arguments into theories. The way they approached the unknown changed the way of thinking of future generations. The over-used phrase 'thinking outside the box' could be truly applied to them.

The omniscient spirit of Chara, the bright girl who lifted me out of a ditch, is following me, ready to pull me down if I climb a hill of flawed knowledge. 'Isn't science just a Greek way at looking things?' she asks.

'You flatter us ancient Greeks, Chara, but let's us not forget that Babylonians, Sumerians, Mesopotamians, Egyptians, Chinese, Hindus and other great ancient civilisations have also contributed greatly to the beginnings of science.'

Anaximander (*c*. 610–*c*. 547 BC), who was my student, boldly declared that the earth was like a disc, suspended freely and without support at the centre of the universe. It remained there because it was at an equal distance from all other heavenly bodies. These heavenly bodies were rings of hollow pipe of different sizes that were placed on circling wheels, which moved at different speeds. This first dynamic model of the cosmos, which could be used to visualise the movements of planets, makes me a proud teacher. Anaximander's view of the cosmos was not only rational but revolutionary: it was conceived at a time when scientific and philosophical thoughts were deep-rooted in our myths and literature.

Chara is laughing and coaxing me to tell you why Anaximander rejected my most important theory. She then charms me by saying that my observation that amber (*elektron* in Greek), if rubbed, would attract feathers and other light objects was indeed the first step towards the discovery of electricity.

What are things made of? I was the first to ask and answer this question. I said that everything in the world – land, air and all living things – had begun as water and would eventually become water once again. What I had proposed was a theory about the origin of things, which competed with the creation myths that were popular during my days. Anaximander rejected my theory because he thought that the thing with which the universe began could not be similar to any of the ordinary things familiar to humans. He said the primary element of all things was an indefinite something called 'boundless' (*apeiron* in Greek). The heavens and the worlds within them arose from *apeiron* which has no beginning, no end, no definite shape or quality of its own.

Chara frowns. 'Weren't his arguments false?'

'Yes, but he pioneered penetrating objective analysis, the corner stone of science.'

Anaximander's way of thinking was the start of the scientific method. He extended the concept of human laws when he said that the physical world was also governed by laws. One of the aims of science is to search these laws of nature. He was indeed the founder of Western science.

Cosmo stopped reading. 'I read it to Mrs Digby and she loved it. She didn't like when I told her that it was not the Gospel, but more believable on many questions.'

'You should not tease my Mama about the Bible. She will throw you out of the house, Cosmo. You know what that means?'

'Losing my job – and you. Whatever you say about your mother, her faith is not blind. She enjoys an argument based on reason.'

'So you think you have turned my Mama into Thomas Hobbes? Like him, she now believes that superstitions have originated in the ignorance of natural causes and have been preserved by crafty men for the purpose of keeping them in power?'

Ignoring my sarcasm, Cosmo gazed lovingly at me. 'You have absolute power over me, Jane. Cool down. Your anger makes my heart flutter.'

I giggled nervously. 'Your heart will flutter wildly if I tell you today's story. The supernatural spell of the Witch of Ware will make you tremble with fear. First, finish reading the scroll.'

Anaximander and I lived in Miletus, a Greek city near the Aegean coast of Turkey. The ancient Greece was not one country – it was a collection of city-states (the *polis*, from which comes 'politics') extending from mainland Greece, to the islands scattered throughout the Aegean Sea and to the Aegean coast of Turkey. Each city-state had a patron god or goddess.

To meet Anaxagoras (500–428 BC) I shall take you to Athens, the largest *polis*; its patron, Athena, goddess of wisdom, but first I must thank Pythagoras for proposing that the earth was spherical. This revolutionary – yes revolutionary, if you look around it's hard to convince yourself that the earth is anything but flat – idea was based on his observation that the earth casts a circular shadow on the moon during eclipses. He offered another argument: the circle is the most perfect geometrical form; the sphere is the most perfect of all solid figures. Hence the earth, the sun, the stars, the planets and the universe as a whole, must be spherical. The idea of a round earth was soon accepted by other philosophers and became common knowledge.

Anaxagoras was twenty-two when he read a poem of Parmenides, a Greek poet and philosopher, which said that 'the moon wanders about the earth, shining at night with borrowed light'. From this Anaxagoras deduced the reason for the eclipses of the sun and the moon. He said that the sun was a red-hot stone and the moon was of earthy nature with valleys and mountains. When the opaque moon came between the earth and the sun, it would cast a shadow on the earth causing the eclipse of the sun. Similarly, when the earth came between the sun and the moon, the earth's shadow caused the eclipse of the moon.

'Didn't you say a theory has to be tested against observations?' Chara asks. 'How did your wunderkind prove his theory?'

'The total solar eclipse of the 478 BC gave him an opportunity to test his theory.'

As the eclipse was caused by the moon's shadow, the shadow would have a finite size. This means that some people would see the eclipse and some would not. Anaxagoras went to the harbour and asked sailors who had travelled around the Peloponnesian peninsula (in southern Greece) whether the eclipse was visible to them. From these enquiries he found that the moon's shadow roughly covered the Peloponnese and people there saw an entire black disc in front of the sun blocking most of the sun's light, but people beyond the Peloponnese only saw a partial eclipse. He concluded that the moon was as large as the Peloponnese and the sun many times larger than the Peloponnese. These figures were wrong, but they were derived scientifically. He was the first practitioner of the scientific method.

Anaxagoras believed he was born to investigate the heavens. Athenians, however, didn't like his investigations. He was accused of teaching that that the sun was not a god but merely a mass of red-hot stone. He was charged with impiety, the lack of reverence for religion, and forced to leave Athens. After the trial someone remarked, 'Anaxagoras, you have lost the Athenians.' He was quick to reply, 'No, they have lost me.' In fact, reason had lost to superstition. Anaxagoras' trial was the first sign of tension between religion and science, which continues to this day.

Clever Chara comments rightly that Anaxagoras was the first man in history who was punished for thinking freely. She says, 'Tell your readers, please Thales, how the superstitious Greeks came up with such rational scientific ideas.'

The ancient Greeks loved beauty and sanctity but also magic and mystery. They attributed their fortune or misfortune to the gods. Their myths, stories of the exploits of gods and heroes, underpinned their spiritual world and were the core of their collective consciousness. While common people were psychologically captivated by gods, as early as the fifth century BC many educated people started recognising that gods may have nothing to do with every illness or storm. They started distinguishing myths from natural causes. Mythology was no longer a hindrance to the search for truth and the love of science. Their imagination was not constrained by religious dogma because they didn't have theology, the systematic study of religious dogma. They employed systematic scepticism to their inquiries.

I rubbed my eyes.

'Did you know, my sleepy girl, there are no scriptures in science, no blind faith; every idea has to be verified through experiments or mathematical proofs?' Cosmo said smugly.

'Science has its own laws and prophets. So what's the difference between science and scriptures?' I snapped back.

Cosmo ignored me.

Aristarchus (310–230 BC), who lived on the island of Samos, was the first Greek scientist to propose that the earth moved around the sun. This idea seemed utterly strange in those ancient days and it was, of course, rejected by his contemporaries. They thought it ridiculous to imagine that anything as big and solid as the earth could be in motion! If the earth did move through space, wouldn't clouds in the sky and birds in flight get left far behind? If you jumped, wouldn't you land in a different spot as the earth travelled beneath your feet?

'Did you feel it moving? I didn't. Tell your readers not to laugh because one day a future generation would laugh at their common sense,' Chara urges me. 'Common sense is nothing but prejudices we collect as we grow up.'

Aristarchus couldn't provide answers to such 'common sense' arguments. Science had to wait for nineteen centuries before Galileo demonstrated that while you are airborne everything around you is carried forward with the earth as it rotates on its axis and moves about the sun.

The charge of impiety was also brought against Aristarchus because he removed the furnace of the universe from the centre and set it into motion. Luckily, nothing happened to him.

Cosmo stopped reading. 'Tell me what happened this morning.' He was extremely anxious while I related him the events of the day. 'Do you know what an imp is?' he asked.

I looked at him with a quivering question on my face. 'Nnnn!'

'An imp or a familiar is a demon set up by a witch to do evil deeds. In return the witch allows the imp to suckle her blood through a lump on her body.'

'I get it. The cat I saw was an imp.'

'You shouldn't say that.'

'Sorry. I shouldn't jump to conclusions.'

'A hundred years ago in Scotland, a witch christened a cat, bound her with joints from a corpse, rowed out to sea and cast her in it. This raised a violent storm and nearly wrecked one of king's ships. For this crime, she was strangled and then burnt at stake.'

A shiver passed through my spine. 'Unbelievable!'

'In England they don't burn witches; they hang them.'

'It's still barbaric. But why?'

'They are vermin to be exterminated because they have switched their allegiance from God to Satan. Sorcery was invented by Satan to lure humanity away from Christianity. That's what people say, not me.'

'Could they hang a woman just because some girls say that she is a witch?'

'No, but at least thousand women have been hanged in the past couple of centuries, mostly on similar flimsy charges. Today's unlucky turn of event has made me worried. Vicar Tayler is a crafty man looking for an issue to show his power. Mrs Digby told me this morning that he is keen to get to the bottom of the girls' claim. Now by asking Mr Bartlett to see him, Dr Digby has opened a door. That chicken-hearted Bull of a landlord would be singing chrr-chrr-chrr to Vicar tomorrow morning. The future is grim and forbidding.'

A cloud of sadness descended over me. 'There was no Satan in Greece and no witches. Please take me back to Greece.'

His name was a synonym for science for two millenniums, but preconceived notions distorted his conclusions.

Aristotle (384–322 BC), an Athenian, is not on my list of the three discoverers of the cosmos. However, it's imperative to look at his cosmos because for two millenniums his ideas ruled scientific thinking and were accepted without question.

'Wasn't he the one who liked to wear fashionable clothes, rings on his fingers, and took considerable pains to dress up his hair?' Chara interrupts me again. Ah girls! Looks are so important to them.

'Why he had to be like an archetypical philosopher like me, unkempt hair, undisciplined beard, crumpled clothes?' I retorted.

Cheeky Chara adds, 'And absentminded.'

'Under Aristotle's ordered hair was a unique inquiring mind infused with investigative spirit.'

Aristotle's cosmos was of finite dimensions, uncreated and indestructible. The spherical earth was at rest at its centre, with the moon, the sun and the planets move around it at uniform speed in invisible concentric spheres. These spheres were bounded by a celestial sphere, on the concave side of which were the fixed stars. A divine force, namely God, spun this celestial sphere on its axis every twenty-four hours.

The stars were spherical and did not have any independent motion. They were not hot; their light and heat came from friction with air. Meteors and comets were not celestial objects because like planets they did not go around the sun in circles; they were related to meteorological phenomena such as lightning and rainbows.

He suggested that everything in the cosmos was made from four elements – earth, water, air and fire. Each of these elements had its own type of movement, in a certain direction: cold and dry earth went down, hot and dry fire went up, cold and wet water went above earth, and hot and wet air went above water but below fire. As the moon, the sun and all the other planets and stars travelled in a circle, different from the straight-line motion of the four elements, he proposed that they were made of a heavenly fifth element, ether, whose natural movement was circular.

Like Anaxagoras and Aristarchus, Aristotle was also charged with impiety. He fled Athens, saying that he would not allow Athens to sin twice against philosophy. He was referring to the trial of the renowned philosopher Socrates, who was accused of corrupting the minds of young Athenians and was sentenced to death by drinking poison.

Aristotle will always be admired as a great philosopher, but in matters of science he was wrong most of the time. Yet there has never been a scientist whose teachings received a kind of divine reverence for so long. Some of his ideas even stalled the progress of science.

'Why did the mighty Aristotle fail as a scientist?' Chara asks.

'To scientists, the theory decides what they observe, but Aristotle's power of observation was distorted by preconceived notions.'

'If you desire to learn, you must learn to doubt. Didn't he say that? Why didn't he doubt his own preconceived notions?'

'This town is full of people with preconceived notions,' Cosmo blurted in anger. 'I fear their preconceived notions would soon turn bloody. How could we stop them, Jane?'

My mind was numb. I whispered. 'I don't know.'

'Who's this Thales? Why's he sending these scrolls to us?'

'Don't know.'

'We must find out, Jane.'

'Don't know.'

'C'mon, Miss Don't-Know, think like a scientist.'

'Don't bore me.'

'If you find the company dull, blame yourself, Miss Bore.'

'Read on, you haven't finished yet.'

'How do you know? Didn't you say you can't read mirror writing?'

Socrates practised induction, a logical process in which a general statement is built from a series of related observations: Thales is mortal; Thales is a man; therefore, all men are mortal. (Hypothesis, a tentative explanation of observed facts, is a stage beyond induction; every theory and law in science begins as a hypothesis.) Aristotle's logic revolved around one notion: deduction (*syllogism*). Deduction is a logical argument consisting of a major premise, a minor premise and a conclusion: all girls love gossiping; Chara is a girl; therefore, Chara loves gossiping.

'Little girls never tell lies; Alice, Agnes and Amy are little girls; therefore, Alice, Agnes and Amy never tell lies.'

'Shh! Jane.'

Science, Aristotle said, is a logical system in which the truths of nature can be deduced from universal principles. These universal principles were perceived intuitively as correct; but, like axioms of geometry such as 'a straight line can be drawn between two points', were unprovable. For example, from the universal principle that the circle was the perfect shape in nature, Aristotle deduced that the heavenly bodies must move in circles. Such was Aristotle's influence that when in 1609 Kepler calculated that planets move around the sun in ellipses not in circles he had difficulty in convincing himself the truth of his own discovery.

'Jane, the universal principle these days is that witches exist. From this universal principle everyone would soon deduce that Mrs Marsh is a witch.'

'How can we nullify the influence of Vicar Tayler, the Aristotle of Hertford? Cosmo, please find some way.'

'Observation, induction and deduction should always be followed by verification. Aristotle failed to verify his results and that's why he failed as a scientist.'

'Perhaps Salus can convince Vicar to verify girls' accusations.'

'Will he or will he not, that's the question.'

Broomstick

A cat might have nine lives, but even in its first life Mrs Marsh's cat appeared in more incarnations than Hindu gods, all within days after Hertford buzzed with the rumours of a witch in nearby Ware. George Gifford, the High Street draper, confided to Vicar that soon after dark a grey cat came through the window of his shop and metamorphosed itself into the person of Elizabeth Marsh. (I doubt he had ever seen Mrs Marsh.) The cat disappeared when he declaimed, 'I defy the Devil and all his works.' About the same time, three miles away in Ware, a monstrous black cat leapt on the shoulders of Dorothy Sampson, wife of the innkeeper of the Crown, as she went down the cellar to draw some beer. The cat's weight, which was supposed to be the weight of Elizabeth Marsh, brought Sampson crashing to the floor. The cat said in a very clear voice that the next time Sampson would come down to draw beer in her cellar she would find bull's blood in the barrel. Shaken badly she couldn't utter a word for a few minutes. When Mr Sampson placed the Bible in his wife's one hand and a glass full of brandy in the other, she narrated the incident after many quick gulps and advised her husband, 'Don't tell Susan Bartlett, she would be worried to death.' Even before she could finish her brandy, Crown's snoopy drinkers had relayed the news to Bull's drinkers. Upon hearing the story, Mr Bartlett immediately ordered the brewer to make his ale and beer safe from bewitching by plunging a red-hot iron into the vats. Not to be outdone by these occurrences, Mama reported that a white cat followed her at dusk but ran away when she entered the parish church. These and many other encounters happening within hours and within a radius of three miles defied all natural laws of gravity and speed of travel of material bodies, but witches are supposed to have supernatural powers, ain't they?

To ward these supernatural powers off and to avoid bad luck, when Malcolm Landish, a baker in Ware, went on a business trip to Hemel Hempstead, he left a few pennies on Mrs Marsh's doorstep; a common practice to please witches. He blabbed that he had watertight evidence that Mrs Marsh was a witch as last week he had seen her at the bank of the River Lea splashing water on her cat and before he got home there was a heavy downpour. Not only this, he said, she shot a baleful glance at him that his limbs ached and trembled. The great pain lasted until he knelt down and prayed – in the pouring rain. Only a witch could do such things to you simply by looking at you venomously; their evil eyes send deadly rays of spirits which would hit you like lightning.

The parish constable's wife Deborah Bowes openly told anyone who cared to listen that witches cut off a piece of their garments and as a token of homage give it to the Devil, and then quickly added that she had seen a piece missing from Mrs Marsh's petticoat. And Mrs Marsh's neighbours had never seen a bottle of salt at her table. Salt never appeared at witches' tables. And Mrs Marsh walked like a witch. Witches have a funny way in their walk – it's more like a wamble than a proper step. And Mrs Marsh never wore red clothes. Apparently witches dislike red colour.

To town folks' surprise Mrs Bowes proved herself worthy of the title of the Hertfordshire's Walking Encyclopedia of Witchcraft Remedies. She advised a terrified Mrs Bartlett to hang leaves of mugwort on all doors and place a needle or a bodkin under a stool to chase away the witch. She told her to make three cuts under a table with a knife and leave the knife stuck there. It would prevent the witch to sit down in the house, in case she managed to get in spite of all other charms. As a further precautionary measure she suggested that Mrs Bartlett burn the hair and nail clippings of her bewitched daughters. This, she assured Mrs Bartlett, would bring the witch into the house in agony. When it didn't happen, she asked her to boil the girls' urine to cause the witch discomfort. This remedy also failed. She also advised everyone not to give any personal items to Mrs Marsh so that she couldn't work her magic through them; and if they had given any item to her in the past they could unwitch themselves by placing three leaves of sage and three leaves of St John's wort in ale and drink it last at night and first in the morning. She encouraged mothers of young children to give their charges sips of dill-water and coral rattles to play since dill and coral were witch-proof. Following her many advices, Mama planted a bay tree near our front door for its anti-witch qualities, and placed a metal horseshoe under the doormat to prevent the witch to cross the threshold. Mama needed special measures because Salus was trying to undo the witch's spells on the girls and therefore a prime target of her spells.

Jennit Hart, a seamstress, suddenly remembered that one day six years ago Mrs Marsh had responded to her little boy's playful taunts about her walnut-like face by smacking him and threatening him, 'You will pay for it, child.' The same night the boy fell sick violently and died within a week. When the only child of Margerie Throckmorton, a relative of Mrs Hart, fell sick, instead of visiting Salus as she had done in the past, she went straight to Mrs Marsh and asked for her forgiveness. The petrified Mrs Marsh placed her hands on the delirious child's burning forehead. The child soon recovered completely. Alarmed at the medical and magical powers of Mrs Marsh, Mama placed a piece of a page of the Bible in an amulet and asked Salus to wear it around his neck. Cosmo and I were amazed when he demurely consented to wearing it.

I wondered whether Hippocrates also wore an amulet when I read about the great physician in the scroll I had found hanging from the door. I didn't tell Salus and Mama about the scroll. Of course, Cosmo had to read it to me. I didn't think I could ever master mirror writing. We were both surprised that this time Thales had addressed the scroll to us.

Hippocrates prescribed the right medicine to ward off evil spirits. It didn't work like his wonder drug; the patient was still superstitious.

Thales again, Jane and Cosmo. I'm sure you have seen the little marble bust of a bearded Greek man that serves as a paperweight on Salus' desk. It's ironic that a statue of the man who is remembered as the 'father of medicine' has now become a mere piece of stone in a physician's house.

Very little is known about Hippocrates (c. 460–377 BC). He was a contemporary of Socrates and lived at Cos, an island in the Aegean Sea.

About sixty of his medical writings have survived and they are known collectively as the Hippocrates corpus. The corpus is the oldest surviving Western scientific text. It laid the foundations of the Western medical tradition by divorcing the supernatural from natural and showing that the disease is due to natural causes and should be treated accordingly. Although its remedies are now considered imaginative, the corpus speaks the language of science; it does not speak of spells, demons or gods.

'Obviously he didn't believe in wearing an amulet to repel evil spirits.' I interrupted Cosmo.

'Goodness knows why Dr Digby does.' Cosmo sniggered and continued reading.

Instead of attributing disease to the will of gods, he introduced clinical medicine: the practical application of intelligent observations. It was a simple yet groundbreaking mechanism. He was also the first practitioner of holistic medicine: the physician must heed the entire patient, considering the patient's whole surroundings. It's the sick person that matters, not the theories of sickness. Health is harmony between you and your environment.

Hippocratic medicine is based on the balance of four elements – water (cold and moist), air (moist and hot), fire (hot and dry) and earth (cold and dry) – and four humours (bodily fluids) – phlegm (water), blood (air), yellow bile (fire) and black bile (earth). Each humour has a typical function: phlegm is a coolant that increases during fevers, blood is the source of vitality, yellow bile helps digestion, and black bile darkens the blood and other bodily secretions. Good physical and mental health depends on maintaining the perfect balance of humours; sickness was the sign of imbalance. For example, fat and lazy persons are afflicted with too much phlegm, and a gloomy state of mind is caused by black bile. The task of the physician is to restore the balance of humours with diet, medicine and exercise.

Hippocrates once said, 'Science and opinion are two things; science is the mother of knowledge but opinion breeds ignorance.' How true!

'You know Thales, once I had pain and fever and my mother gave me extract of willow bark. It worked fast.'

'Hippocrates introduced this wonder drug; now they call it salicylic acid or aspirin.'

Physicians no longer practise Hippocratic medicine, but his name survives in the Hippocratic oath that students take on graduation day in many medical schools. Hippocrates was concerned about physicians' duties rather than their rights, and required his pupils to take an oath to practise medicine according to certain ethical rules. The oath included promises such as 'Into whatever houses I enter, I will go into them for the benefit of the sick'. It warned physicians against overcharging, overdressing and wearing perfume, while urging upon them a decent haircut and trimming of nails.

Cosmo threw the scroll on the table and grumbled, 'Pity Dr Digby's oath doesn't apply to Mrs Marsh.' I ignored him because my ears were more attuned to the voices of Vicar Tayler and Mr Bartlett emanating from Salus' room. As I had never taken any hypocritical oath and didn't intend ever taking one I considered it quite ethical to eavesdrop on their conversation. Conscientious Cosmo was horrified when he saw the ears of his Hippocrates-oath-bound-master's daughter glued to the door between the store room and the consulting room, and immediately pulled me away. Still I managed to hear bits of their conversation:

Mr Bartlett: 'Reverend, sure you know Mrs Bowes, the constable's wife. Yesterday she saw a broomstick with brushy side up outside Mrs Marsh's house. She thinks you would be interested in this sighting.'

Salus: 'Did she also see Mrs Marsh rubbing flying ointment over her body and flying out of her window on the broomstick? A few pious words from Mrs Bowes would have made Mrs Marsh tumble down from the sky. Witches can't stand utterances from the Bible, you know.'

Vicar: 'Don't be sarcastic, John. Mrs Bowes is one of my most devout parishioners. Mr Bartlett, witches usually place a broomstick with brushy side up to ward off evil spirits. That Marsh woman is no doubt is involved in witchcraft.'

Salus: 'I know the broomstick has been connected with witches for centuries, but it doesn't prove that the girls have been bewitched by Mrs Marsh.'

Vicar: 'John, if you can explain their seizures in a medical way, you're welcome to cure them. My good friend, your medical science has failed you simply because the girls do not have any sickness of the body or the mind. You cannot deny biblical facts.'

Mr Bartlett: 'You're absolutely right, Reverend. Did I tell you about the diabolic contraption the missus saw in Mrs Marsh's house? It was like a sundial but divided into seven sections, not twelve. She thinks it was made of bone.'

Vicar: 'It must be a witches' moon-dial which has only seven sections for seven hours of dread. I'm pretty sure the dial Mrs Bartlett saw was made of human bone.'

Mr Bartlett: 'I've never heard of the moon-dial. Sure it's another proof that our neighbour is a witch.'

Vicar: 'Tomorrow we'll go and see that Marsh woman.'

Salus: 'I'm not convinced, but you may talk to Mrs Marsh. The girls are my patients. You shouldn't talk to them without my permission.'

Vicar: 'I agree with you John, you're bound by your medical ethics to care for the girls. I need to talk to Mrs Gardiner about ...'

~~~

Cosmo grumbled, 'In the world of mediocrity, genius is dangerous,' picked up the scroll from the table and started reading again.

This domineering doctor practised for fifteen centuries. No one dared to check his pulse to see whether his out-of-date ideas were still breathing.

If we rank Hippocrates as the first physician worthy of being remembered, then Galen (*c*. AD 129–*c*. 216), a Greek physician who practised in Rome, would certainly be the second top doctor. Yet his influence on future medicine was greater than that of Hippocrates.

He believed that Hippocrates was never wrong – mostly obscure – and saw his work as the extension and explanation of the Hippocratic corpus: 'It is I, and I alone, who have revealed the true path of medicine. It must be admitted that Hippocrates already staked out this path. He prepared the way, but I have made it possible.'

Galen was an overbearing man with the answer to everything. Chara remarks sarcastically. 'A disturbing personality trait in doctors even today.'

His ideas ruled medicine for fifteen centuries. His teachings (though many of them were wrong) became the ultimate medical authority and thus stopped the progress of medicine. As late as 1559 the Royal College of Physicians carpeted an Oxford doctor who dared to question the dictatorship of Galen's dogma.

He started his medical career as a physician to gladiators and later became the physician to the Emperor Marcus Aurelius. As religious restrictions forbade the dissection of human remains, he experimented mainly on apes and pigs. He was the originator of experimental method in medicine. Many of his observations were correct: arteries carry blood, not air (which was common belief); and urine was formed in the kidney, not in the bladder. He pioneered the technique of taking the pulse, still done by doctors to this day.

He was an amazingly prolific writer, with nearly five hundred treatises to his name, of which about one hundred and twenty have survived. He wrote his first book at the age of thirteen and the last in the year of his death. He saw science of medicine as based on two criteria, reason and observation.

In his later life he gradually became a theologian, yet continued to speak with the voice and authority of a scientist. He hated philosophers and scientists who would explain the wonders of creation without reference to God. It's ironic that the man who began his life as a believer in reason and experimental method of scientific discovery ended it as a mystic. It's also ironic that he was blindly followed as a prophet of a new kind of dogmatism for so long.

'A perfect role model for Vicar.' Cosmo the grumbler walked out of the room.

~~~

I was on my way to Ware with Vicar; Mrs Gardiner had asked me to accompany them, as she put it, because of my friendship with the Bartlett girls. Vicar's demeanour suggested that it was his writ to put Mrs Marsh through the wringer. A stocky fleshy man with a ruddy face, he was badly in need of dieting, exercising and shaving. I added the word 'bathing' to the list when I sat next to him in the coach with Mrs Gardiner and Mrs Bowes opposite us. In our cold climate most people avoided bathing and they had two ready excuses for it: the medical excuse that bathing weakened the body, and the religious excuse that even a dirty body would pour out a pleasant fragrance if the soul it inhabited was holy. It would be sacrilegious, at least in Mama's eyes, if the thought that Vicar's soul was not holy had even crossed my mind.

Except being dirt poor, Mrs Marsh perfectly fitted the archetypical image of a witch: an old, wrinkled, pale, bleary eyed, hairy lipped, sullen, superstitious widow with a ragged coat on her back, a cap on her head, a spindle in her hand and a cat by her side. Her small, derelict timber-framed cottage was on a piece of earth worth a few bobs. She had had two husbands, the first of whom died what town gossip had termed suspicious circumstances, and the second, who owned the cottage, had whinged publicly about being hoodwinked into marrying a bitch with bad temper and sharp tongue. When he died a year after making her the sole beneficiary of his last will and testament, the town gossip revved up.

She must have seen the coach approaching, as she was standing outside the door waiting for us. After a perfunctory exchange of pleasantries, she courteously offered us to go into the house before her. Vicar stopped us. He walked to the coach and came back with his Bible. He thumbed through it for a minute and then started reading almost inaudibly. I heard only the last sentence clearly, 'You belong to your father, the Devil, and you want to carry out your father's desire.' Aghast at hearing these words, Mrs Marsh said in a meek, polite voice, 'No, reverend, I do not believe in the Devil. I believe in the Lord as much as you do, sir.' Ignoring her, he thumbed through his Bible again and read, 'Because you have rejected the word of the Lord, he has rejected you.' No one said 'Amen'.

A tearful Mrs Marsh led us through the front door and then into the sitting room. She disappeared suddenly and reappeared after a few minutes with four mugs of ale. Others ignored her when she passed the tray around but I picked up a mug and immediately took a swig. Vicar, Mrs Gardiner and Mrs Bowes gave me dirty looks as if I had committed a cardinal sin. I ignored them and quickly gulped down the ale.

Vicar looked at me as if waiting for me to belch. When I did not oblige him, he started the proceedings (yes, the preliminary process of accusing Mrs Marsh of being a witch).

'Mrs Marsh, sure you know why we're here?'

'No, reverend, I have no inkling of the cause for your visit, but you're welcome.'

'Do you consort with any evil spirits?'

'No, reverend. Why do you ask me this question?'

'Have you made a contract with the Devil?'

Mrs March placed her hand on Vicar's Bible and boldly said, 'I have never made any contract with the Devil.'

Vicar pulled the Bible away and continued his inquisition. 'Why do you hurt daughters of your God-fearing neighbour?'

'I do not hurt them. I feel as much pain as their mother when I see them in seizures.'

'Who do you employ to hurt them?'

'I employ nobody.'

'What creatures do you employ?'

'I employ no creatures, reverend. Why are you falsely accusing me?'

'Three girls have accused you of hurting them.'

'I have done no harm to them.'

'I have heard myself Alice, Agnes and Amy accusing you of hurting them,' Mrs Gardiner spoke in support of Vicar. 'If you forgive them, Vicar will also forgive you. Won't you, Vicar?'

'Yes, God will forgive her,' Vicar said.

'But I have done no harm to them.'

'Let's us go to their house now. If the girls are happy and healthy while you're there, perhaps we will believe you.'

When Mrs Marsh refused to go, Vicar told her in a harsh and loud voice that he had the authority to compel her to accompany him if she won't go willingly. He looked at Mrs Bowes and said, 'You wouldn't want Mrs Bowes's husband to arrest you and drag you to Mr Bartlett's house. Mr Steven Bowes is our parish constable, if you do not know.' Mrs Bowes and Mrs Gardiner nodded. Mrs Marsh looked confused. I took her hand and we all walked across the road.

We were shown to the sitting room by Mrs Bartlett where her three daughters and their friends Joan and Grace Pollard were standing perfectly calm facing the fireplace. When they turned around and saw Mrs Marsh, they all instantaneously fell down on the floor. They were writhing and wriggling in agony like fish freshly taken out of water. Their bellies heaving up and down like blacksmith's bellows. Their eyes closed as if they were blind, their hands and legs stiff like logs. I had never heard about Joan and Grace's attacks of seizures; I made a note to ask Mrs Gardiner.

Mrs Bartlett looked at a dumbfounded Mrs Marsh and started crying; Mrs Bowes stood still with her mouth wide open as if her encyclopedic knowledge of witchcraft had been blown away in a storm; Mrs Gardiner's midwifery license lost its lure; standing closest to Alice, Vicar Tayler fumbled through his Bible and began reciting a psalm, but stopped when he suddenly felt a force pulling one of the hooks of his breeches; and, while he looked at the hook, a yard-long projectile of vomit from Alice's mouth hit his face. Our dear Vicar joined the girls on the floor. Miss Jane Digby, who was trying to absorbs the details of the scene so that she could describe it to Cosmo in the minutest details, forgot that her midwifery license was years away and she would be waiting for her medical degree for an eternity, transformed herself into a true incarnation of Galen and took Vicar's pulse. He was still breathing but his hand was as cold as clay.

~~~

Everything is made of invisible atoms which are always in a whirl in the empty space around them. That's the nature of things, not spirits.

As cold as clay! This simile would have amused Democritus (*c.* 460–*c.* 370 BC). Not because he was interested in earth being cold and dry, as described by Aristotle, but in the answer to a broader question: What was clay or earth made of, or for that matter, pardon the pun, what the matter was made of?

The Greek philosopher Leucippus (*c.* 480–*c.* 420 BC) said that everything was made of tiny particles so small that nothing smaller was conceivable. Democritus, a pupil of Leucippus, adopted and extended his teacher's ideas. Democritus taught that matter was made of void (empty space) and tiny, invisible, indivisible particles which were always in a whirl. He called these particles atoms (*atomos*, meaning 'indivisible' in Greek). As atoms moved about they collided; sometimes they interlocked and held together, sometimes they rebounded from a collision.

He said our universe is just one amongst many: atoms produced innumerable worlds of different sizes; in some worlds there is no sun or moon, in others they are larger than in ours, and others have more than one. His comment that 'some of the worlds have no animals and plants and no water' is the first-ever scientific allusion to life beyond our planet.

When he became blind towards the end of his life, Democritus maintained that what he could see with the 'soul's eye' was truer and more beautiful than things seen with the bodily eyes.

'Didn't he starve himself to death?' Chara asks.

'Sad isn't. When he found that he was likely to die during a festival, which would then deprive his sister of the festivities, he prolonged his life by inhaling the aroma of hot loaves of bread.'

A century later, Epicurus (341–270 BC) advanced the atomic theory of Democritus and combined it with the pleasure ethics of Aristippus, a student of Socrates who believed that sensory enjoyment was the most important thing in life. His followers were called Epicureans and they lived in a garden. The story goes that an inscription on the gate to the garden said: 'Stranger, here you will do well to tarry; here our highest good is pleasure.' Yet the life of community was simple and frugal. Their food and drink was mainly bread and water. Epicureans continued to flourish for another five centuries and exerted a considerable influence over all ancient philosophy and science.

'The universe consists of atoms and the void, everything else is opinion,' Democritus said. Later Greek philosophers found Democritus' opinion threatening as it opposed the comforting idea of a single world that has been designed to accommodate humans. They preferred to believe in their four elements out of which the whole world was created, and Democritus' atomic theory was lost for two thousand years.

In the first century BC Lucretius, a Roman philosopher-poet who was a contemporary of Julius Caesar (the man of brief poetic boast, 'Veni, Vidi, Vici'), wrote an epic poem, *On the Nature of Things*, which kept alive Epicurus' atomism. The main theme of the poem is that behind all natural phenomena lie eternal, unchanging, moving atoms with infinite variety of shapes which can arrange and rearrange themselves into infinite different forms. Nothing can be created out of nothing, he stresses. He explains how atoms form solids, liquids and gases. A solid substance has a vast number of atoms squeezed together, a liquid not so many atoms less tightly packed, and a gas comparatively small number of atoms which are free to move. The same picture is presented by chemists of today.

On the Nature of Things and its scientific thinking were lost to religious dogma during the Middle Ages. When the last surviving copy of the book was discovered in Italy in AD 1417, it re-entered the consciousness of Renaissance scientists. However, it was still attacked for being irreligious. Interestingly, 'superstition derives from ignorance' and 'the mind is born and will die; there is no afterlife; imagining hell is a projection of suffering experienced in this world' are two of the many propositions covered in the book.

To me he was a true scientist. A true scientist peers through dark clouds of superstition towards sunshine that shows the way to reality. It is possible that scientists may never see reality, because science alone will not lead us to the ultimate truth, Jane and Cosmo.

Cosmo banged the scroll at the table. 'The ultimate truth, Mr Thales, is that there are no spirits in this world. Let me ask you something, sir. Democritus' concept of the void leads to an interesting logical argument: if we say there is the void; therefore, the void is nothing; therefore it is not the void. Please explain this paradoxical statement to me, Mr Thales, if you're know-it-all kind of spirit like my spirited friend Jane.'

Surprisingly, the answer appeared in the next scroll. As usual the scroll was hanging from the cold brass knob of the front door. That morning Cosmo had got up quiet early in the morning to see surreptitiously who was leaving the scrolls, but the scroll was already there before he arrived from his house. Apprentices usually lodged with their masters, but Cosmo lived with her widowed mother in the nearby Back Street.

Everything is constantly in change; there is no such thing as change. Is it a paradox? Only bearded men are allowed to answer this question.

Cosmo, your question reminds me of an old joke, 'Why are elephants scared of mice? Because, like my grandmother, they don't like mice in their trunks.' We call it paradoxical that elephants are scared of mice. 'Paradoxical' often means no more than 'odd' or 'absurd'. In science, a paradox is a statement that sounds reasonable but leads to a self-contradictory conclusion. A fallacy, on the other hand, is a statement that leads to a false or absurd result because of improper reasoning.

The argument you have stated is a result of confused mixture of logic and observation. Democritus and his followers simply ignored it. The reason: motion is a fact of experience, and therefore there must be void, however difficult it may be to conceive. I will introduce you to Zeno who said that all motion was illusion.

But first meet two brilliant and highly original Greek thinkers who flourished in the fifth century BC. Parmenides, who gave Anaxagoras the idea of the moon shining at night with borrowed light, and Heraclitus, a native of Ephesus, in Turkey, both studied change. In early Greece students of change were called physicists. The primary meaning of physics is the study of things in motion or change.

Heraclitus believed in universal change: permanence is an illusion and everything in the world is constantly in change. 'You can't step into the same river twice, for fresh waters are ever flowing in upon you,' he said. Parmenides took the opposite position: everything is permanent; there is no such thing as change.

They also discussed the question whether the world of sense-experience is an illusion: Can we trust the senses, or should we rely alone on reason? Heraclitus stressed that senses should be used with caution. 'Eyes are bad witnesses for men if they have souls that cannot understand their language,' he said. Parmenides insisted that reason alone should be trusted as the evidence of sense is unreliable and misleading. 'Do not let habit force you to let wander your heedless eyes or echoing ear or tongue along this road, but judge by reason,' he said.

Zeno (*c*. 490–*c*. 425 BC) was a student and friend of Parmenides and studied with him in Elea. He accepted his teacher's position wholeheartedly and developed further arguments to reject the idea of change. Very little is known about him. However, we know about his paradoxes of motion which say that no motion can be completed because some distance, no matter how small, always remains. These paradoxes had a profound influence on the development of philosophy and mathematics. Their validity is still debated after two millenniums.

The most famous of these paradoxes says that the faster of two runners can never overtake the slower, if the slower is given any head start at all. Thus, Achilles, a hero of the Trojan War reputed to be the fastest runner ever known, will never catch a slow tortoise that started first. Achilles must first reach the point which the tortoise has just left, so that the tortoise must always be some distance ahead. Let's say Achilles can run ten times as fast as a tortoise and the tortoise has a 100-foot head start. When Achilles has run 100 feet the tortoise will have crawled 10 feet, and so will be 10 feet in front. And so it goes on infinitely. Mathematically, Achilles can only keep getting nearer and nearer to the tortoise, but he can never overtake it.

Chara looks at me suspiciously. 'It can't be true. How could a plodding turtle outrace Achilles?'

'It's true as a logical argument known as reductio ad absurdum. You can prove that a proposition is false by showing it leads to illogical conclusion.'

'Could Zeno's argument be used to say that a falling apple is never in motion?

'Yes, at any given moment a falling apple is in a definite position and therefore at rest. Hence it's at rest wherever it is throughout its fall.'

'Could the paradox be resolved?'

'Yes. The paradox is based on the false assumption that space and time are infinitely divisible. It implies the sum of an infinite number of numbers is always infinite. It has now been shown that an infinite number of numbers can add to a finite number. Such a series of numbers is called the convergence series, which occurs when the difference between each number and the one following it becomes smaller throughout the sequence. The race between Achilles and the tortoise is not made up of a number of tiny distances but it is continuous until the end.'

Chara claps loudly like a mad sports fan. 'Achilles has won the race.'

Cosmo stood up and put his hand around my waist. 'Wake up, sleepy girl, and tell me who has been sending these scrolls to us.'

I rubbed my eyes and murmured, 'Thales.'

'Just because someone sends letters in mirror writing doesn't make the person a spirit. A spirit can't interact with matter. It can't hold a pen and write on paper; it can only influence the mind. That's the paradox.'

'It's not a paradox.'

'Okay, I'll tell you a real paradox. Suppose Edy, the High Street barber, puts a new sign on his shop: I shave only and all men in Hertford who do not shave themselves. Does he shave himself?'

'I don't care.'

'You should otherwise you would be marrying a bearded man. If Edy does shave himself, he would be one of the men who shave themselves – meaning he shouldn't shave himself. If he doesn't shave himself and decides to wear a full beard, then he should shave himself according to the sign'

'Well, Edy can end this nonsense by simply taking the sign down.'

'Going back to Thales, there's a catch in it somewhere, absolutely. How do we find the truth?'

'Pray.'

'To whom.'

'The Omniscient One.'

'Thales is definitely not an omniscient spirit. He tells us only what is written in the books. A spirit should know more than that.'

'Why?'

'A spirit is free to go wherever it pleases to go.'

'Yes, but it cannot go forward in time. It's against the laws of nature.'

'Maybe. If I were a spirit I would like to visit Amy Bartlett right now.'

My voice quivered. 'Do you fancy her?'

'Not at all. I'm curious to find out what's going on. She is after all the leader of the gang.'

'The girls' seizures are real. They are not acting.'

'How can I not agree with my charming friend? But this time I'll disagree with your diagnosis, Dr Jane Digby, ma'am.'

'Don't mock me. You know Salus has become edgy and defensive about the girls since Vicar went to see them. He will show you the door if he finds out that you want to talk to them.'

'It'll break my heart. With you, time passes so quickly, but without you, Jane, it would turn from a bird on a wing to a crawling worm.'

I blew a kiss. 'Be nice to me, my bard, and I'll show you where and when you can meet the girls without anyone noticing it.'

He nodded with a yearning look in his eyes. I soon realised that he was not admiring me, but the liquid in the flask on the bench which was slowly turning from white to blue. Cosmo was not interested in the union of bodies or souls, but in the union of chemicals. I needed a cauldron to cook a magical potion to transform Cosmo the amateur alchemist and would-be physician into a lover, but I doubted whether Mrs Marsh would have one in her house. Well, Mrs Bowes hadn't told us about it, yet. If she did report about a cauldron in Mrs Marsh's house, would she say, like that balding bard, it was bubbling with blood of bat, juice of toad, scale of dragon, tooth of wolf, finger of birth-strangled baby, gall of goat, maw and gut of salt-sea shark, root of hemlock dug in the dark and slips of yew tree slivered in the moon's eclipse?

Demons

I had heard that constable Bowes, a snub-nosed, languid man, was a bit of a sissy. A daydreamer, he was more interested in tending his collection of bawdy and lewd books, most of them in French, than carrying out his sworn duties. His collection even included the incredibly saucy and raunchy *L'escholle des Filles*, I had been reliably told. He spent most of his time in public alehouses. As pub brawls were common, 'the maintenance of the king's peace' was his ruse for visiting them so frequently.

You are absolutely right if you ask: how come a nice girl like me, especially who hates gossiping, knew so much trivia? You see I used to spend a lot of time in the store room overhearing patients gossiping in the adjoining room while they waited for their turn. You want me to sing the John Blow anthem 'O Lord, I have sinned'? Nope. No Hail Marys either. In my book, eavesdropping is not a sin; gossiping is. Why? Eavesdropping is a passive event; gossiping an active one.

A patient once confided to another patient that he had overheard Mr Bowes saying it loudly after having one too many at the Elephant that 'all witchcraft comes from carnal lust, which in women is insatiable'. You have got to trust the opinion of a scholar of porno books on the delicate matters of sex. From his comment, Cosmo, my Socrates, sharply induced that the snub-nosed swine would tow his wife's line in the matters of witchcraft. Not good news for Mrs Marsh.

Though there was some good news for Mrs Marsh when Mr Bowes refused to go along with Vicar's idea of giving Mrs Marsh a swimming test. The test followed from the belief that a witch would not sink since the pure element of water rejected those who had renounced their Christian baptism (and then had taken part in a kind of sacrilegious baptism by the Devil). Lazy he might have been but our constable was not naive. He advised Vicar that the test was illegal: to swim a witch was to assault her. If Mrs Marsh died, they could be charged with murder. Not that long ago the swimming test was acceptable: in sixteen-thirteen in Bedfordshire a rope was attached to the waist of a suspected witch and she was thrown into a pond (the idea of attaching a rope was to save the innocent from drowning). The knowledge of the new law on the part of Mr Bowes saved Mrs Marsh from being tied by a rope and thrown into the River Lea. My cynical friend Cosmo pondered for a while on the empty idea whether Mr Bowes read a porno book wrapped in a law book or was it the other way around. My answer was that Mr Bowes had never read a law book in his life but had overheard about the statute of the swimming test in a court session. Eavesdropping has its benefits.

On the matter of the common practice of scratching the witch, the law was silent. The bewitched person scratched the face or arms of the witch with the fingernails until blood was drawn. Scratching didn't prove whether a witch was guilty, but it did destroy the witch's power. The aim of scratching was to cure the bewitched by breaking the spell of the witch. Mr Bowes advised Vicar to go ahead if he could persuade Mrs Marsh.

I found myself again on Vicar's coach to Ware in the company of Mrs Gardiner and Mrs Bowes. This time I made sure that Mrs Bowes sat next to the man of the cloth and soaked up sweet-smelling scents and scintillating spiritual stories. On the way Mrs Gardiner told us that Mrs Pollard and her daughters would also be present at the Bartlett house and it would be my duty to persuade Mrs Marsh to come to the house. It pleased me mightily that I was now in their eyes a little more than a slip of a girl.

When we arrived at the house, we were cast adrift in a sea of bizarre events. Mr Bartlett and Mr Pollard were standing in the porch and looking as forlorn as a spurned teenage girl's diary. I noticed Mr Pollard's balding pate when he doffed his hat. He was dressed to kill: in fine coat, waistcoat, breeches, gloves and high-heeled boots tied with a red ribbon. A handsome, athletic, balding, virile man on his way to see his mistress, I let my imagination run wild for my dear diary. Mr Bartlett came forward, bowed and pointed us to go inside without saying a word.

Inside the hall the scene turned out to be surreal and all too real. Mrs Bartlett and Mrs Pollard, a plain mousy woman of about thirty-five, were alone in the hall. Their eyes aloof, faces grim, bodies trembling. Like their husbands they didn't say a word and pointed us to go upstairs. I went first as I knew the girls' bedrooms, other followed me. But when Vicar put his foot on the stairs, Mrs Bartlett came running and said, 'Sorry Vicar, you shouldn't go upstairs as the girls may not be appropriately dressed.' I knew immediately it was her way of saying that seeing a man dressed in a cassock and surplice would make the girls go out of their heads.

They were already out of their heads – and probably out of their bodies as well – for they were standing in a circle with their bodies as still as the gnomes in the garden outside; their arms outstretched with each girl's taut left-hand forefinger almost touching the flaccid right-hand finger of her neighbour, like the hand of God giving life to Adam in the Michelangelo fresco 'Creation of Adam' in the Sistine Chapel. They were saying things in low, tranquil voices. Their behaviour was diametrically opposite to what I had seen previously. They seemed to be in some kind of trance, supernatural trance.

I didn't know about Mrs Gardiner and Mrs Bowes – their mouths were still wide open in amazement – but I soon worked out that the girls were talking to each other. Rather five spirits, Tig (Amy), Bik (Agnes), Riv (Alice), Nim (Joan) and Pif (Grace), were talking to each other.

Pif: 'What are we doing here, Tig? Let's us go back to our world where there is no light, no darkness, no high, no low, no up, no down, no today, no tomorrow, no beginning, no end.'

Riv: 'Pif is right. Let's us go back to where there is no virtue, no vice, no love, no hate, no union, no separation, no life, no death.'

Bik: 'Riv, we can't leave these girls alone in a world where joy is brief, sorrow endless.'

Nim: 'But whose fault is this?'

Bik: 'Don't you know, Nim?'

Tig: 'There is a witch in this town. From the sole of the foot to the crown of the head she is smeared in the Devil's potion. When she has gone sorrow will be brief and joy endless. Only then we shall leave.'

I knew many spirited people but I had never had the opportunity to engage myself in a lively debate with spirits. I had to have a go.

'Dear Tig, Bik, Riv, Nim and Pif, tell me what are you?' I chimed. 'I can't see you, the words you speak come from the mouth of girls.'

There was a long silence before Tig replied. 'You live in the realm of the senses – the eye, the ear, the nose and so forth; we live in the realm of the mind. Your senses are ignorant of us. If there be ignorance, then knowledge must die. You have no knowledge of us.'

I had to admire the crispy answer coming from a girl, er spirit, of my age. For a moment I thought, my idea of becoming a licensed midwife was not that promising and I should ask her how one could become a licensed spirit. I decided to be diplomatic. 'You have made my ignorance disappear, Tig. But where do you live?'

'Our dwelling is beyond your reality.'

'If you are a forgiving spirit, which you're, would you let the girls scratch Mrs Marsh? Scratching would destroy Mrs Marsh's bewitching powers.'

'The witch must die,' Riv replied.

'A witch's powers are destroyed only when she dies,' Bik added.

'Bik, we'll will let the girls scratch the witch,' Tig announced her decision in a firm voice. 'Next time come with a man of law, not a man of God. Prayers torment the bewitched. Don't you know, ignorant young lady?'

'Thanks.' Feeling flustered that's the only word I could utter.

'Can we leave now, Tig, please?' Pif asked. Pif seemed not to be enjoying the experience of inhabiting the body of Grace.

The moment Tig said 'yes' the girls fell on the floor. They were calm, not in a fit. I checked Amy's pulse. It was normal. She seemed to be in deep sleep.

~~~

When I caught up with Cosmo he was on his way to a patient's house near the Cowe Bridge to deliver medicines; from such errands he made a few pretty pennies. I walked along with him, Nyx ran ahead of us. Nyx was his dog, as black as night.

'Are you going to that huge honey-coloured stone house?' I asked.

He walked quietly until we reached High Street and then replied, 'Yes, Mr Johnson's house. He made tons of money from bribery while he was the clerk of works for a big building project, a cathedral in London, I think.'

'Bribery?' I exclaimed.

'Yes, bribery is as English as mushy peas, a part of our nosy, whinging lives.'

'How does one bribe?'

'If you were a quarry owner and you wanted to sell stones to Mr Johnson, you wouldn't take a wagon load to his work site and ask, Sir, would you like to buy some stones?, you would take a small brown bag full of coins and discreetly place it on his table and ask, Where should I unload my five wagons, sir? Did you know I'm always open to bribing?'

'Why would I bribe you, Mr Ferret? I wouldn't mind bribing Amy, but I have only five pennies in my purse.'

'These little demons would need more than a few lollies to wean them away from their adventure. I should tell you about Mr Johnson's recent adventure. In May he went to Oxford to see his brother. There he saw a neat, new building that had just opened its doors to the public. It was called Ashmole's Museum and he had to pay sixpence to get in.'

'What's a museum?'

'Mr Johnson says that it's a place like a library. Instead of books it has man-made and natural specimens from every corner of the world. But no spirits.'

'Who told you about spirits?'

'Spirits are new bearers of news here, so it travels faster.'

'Mocking me, eh.'

'Mrs Gardiner and our friend Thales.'

'Thales?'

He showed me the scroll. 'I'll read it in a minute. Mrs Gardiner was full of praise for your handling of the spirits.'

'Cosmo, I'm sure the girls are not acting. They either have some extraordinary mental affliction or they are bewitched.'

'Listen, gullible girl,' he scoffed at me, 'there are no spirits and no witches in the world I inhabit. I do not know about your world. Listen to Thales.'

Two men who studied animals and plants and started the science of zoology and botany to keep Men and Women of Scientific Ambitions fascinated.

Though I'm myself a spirit, I wouldn't spend an iota of my time on discussing them. They are nothing but figments of imagination, my friends. Only living things are real. Let's us talk about scientists who pioneered the scientific study of living things, which we divide into two groups, animals and plants.

Our friend Aristotle's most successful scientific writings were about animals. He was a meticulous observer and studied first-hand more than five hundred species of animals, dissecting nearly fifty of them. He discussed human and animal anatomy and the function of the body's organs (though he mistakenly suggested the heart to be the site of consciousness, not the brain); described how chicks developed inside eggs, and baby mammals inside their mothers; distinguished whales and dolphins from fish; studied the social organisation of bees and said that there was only one ruler in the hive (though he called it the 'king' not the 'queen'); observed the metamorphosis of insects ('butterfly is generated from caterpillars which gown on green leaves'); correctly recognised that dolphins were air-breathing animals and not fish, as people believed; and attempted to classify animals into genera and species according to features they had in common – his books on animals are filled with hundreds of these observations, some of which weren't confirmed until many centuries later.

Aristotle devoted himself to zoology as an old man's recreation, yet spent time on observations often continuing for months. No wonder his work on animals was the greatest such collection of the time. It must be said that like his other scientific works it wasn't without errors.

'Zoology began in an old man's hobby,' remarks Chara. 'What about botany?

'It began with the curiosities of a young mind like yours, Chara.'

Theophrastus (c. 372–c. 287 BC) was the first to ask questions such as 'What are the similarities and differences between plants?' and seek scientific answers to them. He was a friend of Aristotle and succeeded him as the head of Lyceum, a school started by Aristotle in Athens.

Theophrastus divided the plant world into two kingdoms, the flowering and the flowerless, and classified more than five hundred species of plants into one of four groups: trees, shrubs, under shrubs and herbs. He also coined new terms to describe their structures and functions, and distinguished various modes of regeneration, spontaneous, from a seed, from a root or from other plants. His botanical works are crowded with crisp and concise observations, proving that he was an acute observer of the life histories of plants.

He was first true botanist and influenced botanists for centuries to come. He was also the first ecologist. He looked at plants in the context of their relationship to the environment, to sunshine, soil, climate, water and other plants and animals.

Theophrastus is remembered now not so much for his botanical works, as for satirical character sketches of various types of individuals – such as the Officious Man ('promises things beyond power'), the Boastful Man ('boasts how he served with Alexander the Great'), the Man of Petty Ambition ('anxious to sit next to the patron'), and the Superstitious Man ('if a cat runs across his path, he will not go on until someone has crossed the path') – which can still be recognised today.

Theophrastus was not a superstitious man and wrote against superstitions such as black cats cause bad luck. Shunning the norms of the society is never a good idea. He was charged of impiety from superstitious Athenians, like Aristotle before him. However, unlike Aristotle, he was cleared of the charge.

Theophrastus, undoubtedly, was the Botanical Man ('studies plants scientifically'), and his teacher Aristotle the Zoological Man ('studies animals scientifically'). If you meet them at a party, wouldn't you be anxious to sit next to them? Most of us are Men and Women of Petty Ambitions.

'Yes, we're. Our not-so-petty ambition is to save Mrs Marsh from being charged of witchcraft,' Cosmo shouted.

'Let the fight against superstition begin,' I joined him.

We were in front of Mr Johnson's house. Mrs Johnson was talking to a neighbour. Both were startled to hear our loud voices. 'What's going on?' Mrs Johnson came forward. 'You alright?'

'Yes,' we both said sheepishly.

'A ferret ran over my feet and it frightened me,' clever Cosmo came up with an excuse to calm down the little old ladies. 'I'm so sorry, Mrs Johnson.'

Mr Johnson looked at her neighbour. 'I wonder who has pet ferrets in our street.' She then patted Cosmo's back. 'You're a good lad. Come on in, bring along your friend.'

~~~

When I didn't hear from Mrs Gardiner about Vicar's next move for many days, I reluctantly asked Mama. She rubbed her hands joyfully. 'Oh, Christmas is only four weeks away and my daughter is asking me about Vicar. He is busy preparing for Advent Sunday festivities, but he hasn't forgotten about the witch. He is going next week. He doesn't care about what the spirits say.'

'Voices of little devils, that's all they are,' I said.

'Vicar Tyler has the power to face the Devil,' she said admiringly.

Leaving her to visualise the great fight between the Devil (assisted by a witch and five bewitched girls) and Vicar (assisted by her, Mrs Gardiner, Mrs Bowes and other like-minded godly women of Hertford) on the meadows of River Lea, I went to the store room where Cosmo was busy grinding some herbs.

'Did Amy the ring leader say that for scratching you have to come back with a man of law and not with a man of God?' he asked.

'Yes, but it was not Amy who said this,' I replied. 'It was Tig the spirit.'

'It's the same thing. Asking for the parish constable is a clever move by her. She wants to entangle Mrs Marsh in a legal trap. She is vicious.'

'I still find your hypothesis that the girls are acting hard to believe. What's their motive? There is a motive behind every crime.'

'Don't you worry; I'm collecting evidence for my hypothesis. If the case ever goes to the court, I'll present the evidence to the court.'

'Okay, tell me your first postulate'. When he remained silent I insisted, 'You must have a postulate to form a hypothesis.'

'It's not the time to tell it,' he replied after a long silence.

'You don't trust me?'

He pretended as if he hadn't heard me and poured the crushed herb in a bottle. 'You know who wrote the most beautiful and enduring postulates?'

'Euclid, of course.'

'Let's ask Thales to talk about Euclid in his next scroll.'

'You must be joking. How can we talk to a spirit?'

'You're an expert in talking to them, my beautiful Jin.'

~~~

We both were surprised when we opened the next scroll from Thales.

The proof of the pudding may be in the eating, but without assumptions there is no proof. If my assumption is 1 = 2, what can you prove?

Greetings Jane and Cosmo. You wanted me to talk about Euclid (325–265 BC), so let us talk about the man who wrote the most famous geometry book of all time, *Elements*. It has been described as a book to be read in bed or holidays, a book as difficult as a detective story to lay down when once began. 'By an old fuddy mathematician like you,' says Chara, 'who has probably never read a romantic novel in his life.' Our Chara can be caustic.

Euclid studied in Athens and then taught at the shrine of the Muses at Alexandria in Egypt, built by Ptolemy Soter, the first king of the Ptolemaic dynasty. The majestic marble buildings of the shrine were crowded with sculptures and pictures and surrounded with a piazza where scholars might walk and converse together. The shrine contained a library which over the centuries amassed as many as 700,000 papyrus scrolls. No other ancient institution has contributed more to the advancement of science and scholarship than the Great Library of Alexandria.

Virtually nothing is known about Euclid's life. 'Don't say that,' Tara admonishes me. 'Everyone knows about two anecdotes about him.'

When Ptolemy asked the great geometer if there was an easier way to learn geometry than by studying all the theorems, he admonished the king, 'There is no royal road to geometry'. According to another anecdote, one of his students complained, as students often do, that learning geometry was pointless – it had no practical value. Euclid ordered a slave to give the student a coin so that he could make a profit in studying geometry.

Elements introduces an innovative logical method of remarkable clarity: from precise definitions, to assumptions to conclusions. A definition ensures mutual understanding of all words. Assumptions are based on explicit postulates (also called axioms). A postulate is self-evident truth that doesn't require proof. It is the basis for an argument. A proof is a series of statements in which each statement is derived from a previously proved statement or a postulate. The final statement is known as a theorem. It is a proven proposition, which is a statement with logical constraints. Sometimes QED (abbreviation for the Latin *quod erat demonstrandum*, meaning 'which was to be shown') is written to denote the end of a proof.

A proof is like a high tower built from blocks of stone. If one block breaks, the tower will topple. Even an innocent false notion in a proof will make it wrong.

Chara challenges me that if I allow 1 = 2, she can prove that she is the Pope. 'The Pope and I are two, but two is one,' she says. 'Therefore, the Pope and I are one, or I'm the Pope.' No wonder, because of its logical excellence, after the Bible, *Elements* is the most translated, published and studied of all the books in the Western world.

Elements begins with twenty-three definitions (such as point, line, circle and right angle), five postulates and five 'common notions'. From these foundations Euclid proved four hundred and sixty-five theorems. Five postulates form the basis of Euclidean geometry: (1) A straight line can be drawn between any two points. (2) A straight line can be extended indefinitely. (3) A circle can be drawn with any given centre and radius. (4) All right angles are equal to one another. (5) If two lines are drawn which intersect a third in such a way that the sum of the inner angles on one side is less than two right angles, then the two lines must inevitably meet; that is, they cannot be parallel to each other.

Euclid's 'common notions' are not about geometry; they are elegant assertions of logic: (1) Things which are equal to the same thing are also equal to one other. (2) If equals are added to equals, the wholes are equal. (3) If equals are subtracted from equals, the remainders are equal. (4) Things which coincide with one another are equal to one another. (5) The whole is greater than the part.

Cosmo stood up and walked out of the room. 'I haven't finished reading yet. Let's go outside. It's too stuffy here.'

I followed him. 'Smelly too. You're always grinding, mixing or heating something.'

We walked towards the old castle. I was in a chirpy mood, zigzagging, twittering.

'Jane, you got to learn something from Euclid. You have never kept to a straight line all your life. You wobble too much in your ideas.'

'What do you mean, Mr Straight Arrow?'

'You either believe in witchcraft or you do not. Do not sit on the fence like Dr Digby.'

'Postulate: There's no Satan in our happy Christian world. Proof: No Satan; no witches. No witches; no bewitched girls. Girls in seizures? Suffer from some unexplained delusion or pretending. Dr Cosmo says they do not have a medical condition. Therefore, the girls are great pretenders, con artists. QED. Happy?'

'It's not the time to joke. It worries me that Mrs Marsh's life could end like that of Archimedes. Let me read the rest of the scroll. '

He was the greatest mathematician, scientist and mechanical genius of the antiquity, but didn't shout 'Eureka!' when he tried to square a circle.

Today Archimedes (*c.* 287–212 BC), the brightest star in the constellation of ancient Greek scientists, is mostly remembered for the tale of his running naked through the streets shouting, 'Eureka! Eureka!'.

Archimedes' legendary bath gave science the buoyancy principle – when a body is immersed it is pushed by an upward force equal to the weight of the fluid it displaces – and scientists a word with which to hail their discoveries – 'and an excuse,' Chara says with a wink, 'to run naked through the streets.'

Squaring the circle – to draw a square equal in area to a given circle – is one of the greatest problems in classical geometry. Almost every great Greek mathematician tried in vain to solve a problem which was, in fact, impossible. Pi, the ratio of a circle's circumference to its diameter, is an irrational number (the value of pi is 3.14159...; Archimedes was the first to approximate this value to 31.14). This means that it could not be written as a fraction or terminating decimal or as an infinite recurring decimal. The ancient Greeks were not aware of this fact. Archimedes was the first to show that the problem is equivalent to finding the area of a right-angled triangle whose sides are equal respectively to the circumference of the circle and the radius of the circle. Half the ratio of these two lines is equal to pi. Archimedes other mathematical works were on geometry of the circle, spiral, sphere, cylinder, conoid and spheroid.

Archimedes, the mechanical genius, arrived at the principle of the lever – used with an appropriate fulcrum or pivot, the lever magnifies force and speed – by reasoning that two equal weights suspended at the ends of a uniform rod, which is suspended at its centre, will balance each other. He also invented the pulley (a type of lever), and the screw, the most widely used basic machine. The lever, the screw, the pulley, the wheel and axle, the pulley, the ramp and the wedge are the six types of basic machines which all form part of complex machines.

Legend has it that when he discovered the lever's incredible ability to make heavy lifting easier he was so excited by the discovery he boasted to King Hieron of Syracuse (a Greek city in Sicily), his relative and friend, 'Give me a place to stand on, and I'll move the earth.' Hieron begged Archimedes to show him some great weight moved by a slight force. Archimedes, therefore, tied ropes to the largest ship of the royal fleet, which had been dragged ashore by the great labours of many men, and after putting on board many passengers and customary freight, he seated himself at a distance from the ship, and without great effort, but quietly setting in motion with his hand a system of compound pulleys, drew the ship towards him smoothly and evenly, as though it were gliding through the water. The crowd was spellbound to see a mortal single-handedly lift a fully loaded ship.

Hieron was amazed by this demonstration and persuaded Archimedes to prepare for him all kinds of offensive and defensive engines to be used in every kind of siege warfare. When the Romans attacked Syracuse, they were panicky to see huge catapults, enormous grappling irons and cranes and big concave metallic mirrors to set fire to the wooden ships. On seeing the destruction caused by a single man the Roman general Marcellus said, 'Let us stop fighting against this geometrical Briareus, who outdoes the hundred-handed monster of mythology.'

When Syracuse was captured, Marcellus gave orders that Archimedes was to be left unharmed. But the orders never reached Roman soldiers who found him in his study drawing some complicated geometrical figures. Seeing the soldiers, Archimedes shouted, 'Do not touch my drawings!' One of the soldiers drove a spear through the body of the greatest scientist, mathematician and mechanical genius of antiquity, when he could well have been contemplating something further for the benefit of mankind. Marcellus was both angered and aggrieved by his death. He buried him with honours and marked his tomb by a sphere packed into a cylinder. Archimedes had asked his friends that at his death his grave should be marked with this particular drawing, with a mathematical formula for comparing their volumes.

'There is a Briareus in our town, you know,' Cosmo grumped at me.

'I'm sure my Marcellus will capture him,' I replied, 'with a plan as precise as Euclid's geometry and as ingenious and effective as Archimedes' machines. Please do not kill him. I can't stand blood. Just wave your magic wand and make him disappear from our town.'

Wand

Advent Sunday is a kind of day when one expects to experience peacefulness and spirituality, but it didn't turn out to be that way, at least for me. In spite of nagging noises, alternating from Mama and Martha and percolating through my pillow, I got up late and arrived at the church after the service had started. The church was full and I could only squeeze into the last pew. When the choir started singing 'O come, Thou Rod of Jesse, free/ Thine own from Satan's tyranny', a hand crept from the back of the pew and gently pulled my bodice. For a moment I thought the hand of Jesus was about to free me from Satan's tyranny, but when I heard someone whisper 'If you were not in the church you'd have a good thrashing, lad' I turned around and saw old Mr Gifford, draped in the Sunday best of his shop, trying to save a damsel from the paws of a prowling witch's cat. The cat this time did not metamorphose itself into the person of Mrs Marsh but my very own Cosmo. Before Mr Gifford could proclaim 'I defy the Devil and all his works' and punch Cosmo, I whispered, 'Mr Gifford, this is James, my father's apprentice.'

Mr Gifford looked pale and puffed. 'Fresh air shall do me good,' he mumbled to himself and shuffled towards the door.

Once outside the nave he rolled his eyes and laughed. 'My goodness, Miss Jane, he is indeed James.' He took a deep breath. 'Since I saw a cat in my shop turning itself into that witch, I'm all nerves these days. My heart palpitates at the slightest provocation. Dr Digby has asked me to take things easy.' He turned up his nose at Cosmo. 'Why are you dressed so shabbily, lad?'

I buttered Mr Gifford up. 'Matter of shame when we have the best draper in these parts.'

Mr Gifford nodded approvingly. 'You're absolutely right. We carry the best lines and at the cheapest prices.'

The disapproving look on Cosmo's face didn't deter me from spreading another layer of cool butter on Mr Gifford and hot chilly on Cosmo. 'My father pays well his apprentices, sir. I'll press upon him to make sure that they are always dressed like gentlemen.' Salus had only one apprentice, but plural always sounds better.

'I'll be glad to see you in my shop, lad.' He shuffled back through the door. 'You coming in?'

'Yes sir, we'll be there in a minute,' I replied. Once Mr Gifford was out of sight, I gave Cosmo a buss on cheek and asked, 'What's wrong with you, silly moose?'

'I have just walked back from Ware. It has been drizzling all the way.'

My heart fluttered like a frightened bird and knees trembled like wind-blown reeds at the thought of Cosmo going to Ware to meet Amy. I could only utter, 'Ware?'

'You haven't heard about Mr Pollard going around the town and collecting depositions against Mrs Marsh?'

'Depositions?'

'Sworn statements from folks testifying against Mrs Marsh.'

'Marsh?'

'I suppose once he has collected a few he will take them to Sir Henry Gilston to convince him that he should issue a warrant for her arrest.'

'Arrest?'

'Sir Henry is the local justice of the peace.'

'Peace.'

'Crikey! The girl who could make conversation with a statue has lost her speech.'

'Na-na-na.'

'I know why you are so shaky. I swear I didn't go to Ware to see Amy. I was there to see Martha's sister.'

I felt like a rock had been lifted from my chest. 'Is she ill?'

'No. Martha told me that her sister Margaret had once seen Mrs Marsh casting a witch's circle and calling spirits to dance with her.'

'Weird stuff.'

'I went to her house to persuade her not to give any depositions to Mr Pollard, but it was too late. She had already done so.'

'Do you know what she said in her deposition?'

'Yes. At Easter Margaret's husband George lost some gold coins. He told his wife to go to Mrs Marsh and ask her to look into her crystal ball and see if anyone has found the coins.'

'Why?'

'Rumour has it that Mrs Marsh dabbles in clairvoyance. So Margaret went to her house. Mrs Marsh welcomed her and immediately told her that she had come to find out who has got her husband's gold coins. She demanded six shilling to tell her the name of the person. Margaret gave her the money.

... She puts on her spectacles and opens a book. She places a round green glass on a page with a picture of the Devil, half man half goat with horns, tail, cloven feet, claws and a pitchfork in his left hand. She says things in some strange tongue, '... te yon kathe mudaren ...', rubs the glass with her hand and then shouts, 'I charge thee Baell to show me true vision in this glass, if any treasures be hidden.' A figure appears in the glass. It has three heads, the first like a toad, the second like a man and the third like a cat. She moves her hands above the glass and the picture changes. It now shows many men and women sitting in a parlour with beer glasses in their hands. One woman is standing near the door with her hands hidden under her apron ...

Mrs Marsh then told Margaret that the coins had been stolen by a brown-haired skinny woman with long face and hollow cheeks. The description fitted with Ellen, a cousin of George. When George confronted Ellen, she confessed of stealing the coins.'

'Do you think it's true?'

'Listen my girl, it doesn't matter what I think, but what Martha thinks.'

'Aeee.'

'After the incident Margaret thought that Mrs Marsh was either a woman of God or a witch. Now comes the real weird part of the story. The day after Ellen's confession George fell ill. Margaret suspected that Ellen had put a curse on him. She went to see Mrs Marsh and asked her to undo the curse. Mrs Marsh again asked for six shillings and when she had been paid, she went outside the house.

.... Wearing only a black linen gown, naked feet, her red hair flowing in the wind, she places a brass pan on a tripod and lights a bundle of twigs under the pan. She then lights four black candles representing the four ancient elements of air, water, fire and earth and places them facing east, west, south and north. She now walks clockwise drawing a circle around the pan with a long stick from a willow tree, her witch's wand, saying, 'I, who am the servant of the Satan, sanctify unto myself the circumference of nine feet round about me, from the east Amaymon, king of the east; from the south Gorson, king of the south, from the west Goap, king of the west, from the north Zimimar, king of the north, which ground I take for my proper defence from all spirits, that they may have no power over my soul or body, nor come beyond these limitations, but answer truly, being summoned, without daring to transgress their bound.' She throws black powder on the hot pan; a thick white smoke blankets the circle. 'Tig, Bik, Riv, Nim, Pif gab gabor agaba; arise, arise, arise, I charge and command thee,' says she waving her wand. The smoke disappears; a black cat jumps into the circle and starts dancing on her hind legs. A knot of toads arrives with a croak, an unkindness of ravens alights with a shout, a troupe of shadows of naked girls appears with a melody. They all dance outside the circle until she throws some white seeds on the ground. They pick up the seeds and vanish ...

Mrs Marsh returned after a while and told Margaret that the curse of George's cousin had been broken. When Margaret came back home she found George as healthy as apples. She is sure that Mrs Marsh is a witch. Her deposition is about these two crazy stories.'

'Do you know about other depositions?'

'Yes, Mr Waterton's. When one of his cows became sick and mooed in pain the whole night, he immediately thought it had been bewitched. Next morning he tried an old charm he had learned from her mother.

.... He fills a bottle with cow's urine and places it in a fire. The urine rises up in bubbles. He stares at the bubbles looking for clues to his cow's illness. As the steam rises up, he sees it shaping into an old woman's face. He goes to Mrs Marsh's house, gives her a few pennies and asks her to come to his cowshed. Mrs Marsh touches the sick cow and sings:

A loaf in my lap,
my penny in my purse.
You are never the better,
and I am never the worse.

The cow gets up, moos happily and starts eating hay from the trough ...

The charm did work.'

'I don't know what makes people to imagine such tales? If Mr Pollard spends a bit more time you could compile an encyclopedia of silly stories, Cosmo.'

He gave me a cold stare. People were now coming out of the nave. We walked out of the porch into the courtyard.

~~~

We both were standing in front of the house when Cosmo noticed a scroll hanging from the door knob. 'Where in the hell has this come from?' he exclaimed. 'It wasn't here when I came in a few minutes before. Thales' spirit lives in a very fast toad.' He seemed upset when he opened it and started reading it.

The first scientist in a Roman toga, the first 'green' scientist, the first encyclopaedist and the first person to die in the cause of scientific investigation.

Cosmo, why waste your time collecting superstitious stories, when you can spend it collecting scientific facts like Pliny the Elder – and with the same dedication?

The Roman statesman and scholar Gaius Plinius Secundus (AD 23–79), known as Pliny the Elder, wrote about seventy-five books on history, grammar, rhetoric and natural history. His only surviving book *Natural History* is the most comprehensive survey of scientific learning in the ancient world. In thirty-seven volumes it covers about twenty thousand noteworthy facts on cosmology, astronomy, meteorology, geography, zoology, botany, agriculture, medicine, human physiology, anthropology, rock, minerals and metals, making it the first encyclopedia of science. In fact, the first encyclopedia of any kind.

Pliny describes the subject of his book as 'nature, that is, life'. He admits that it contains little original work and is a compilation of facts from two thousand other authors, but he assimilates their ideas in such a way as to project an image that was uniquely his own. Written in Latin, this great encyclopedia of knowledge was frequently copied by hand throughout the centuries. Printed in 1469 in Venice, it was one of the first books to be printed in Italy. It was translated into English in 1601.

Natural History ends with the sentence, 'Greetings, Nature, mother of all creation, show me your favour in that I alone of Rome's citizens have praised you in all your aspects.' Pliny the Elder made the science Roman, which until then had been the preserve of the Greeks. 'He also made science green,' Chara reminds me, 'by rejecting reckless exploitation of the earth and encouraging a harmonious partnership between humans and their environment.'

When Mount Vesuvius blew its top in 79 and buried the cities of Pompeii and Herculaneum, Pliny was in command of a Roman fleet stationed in the Bay of Naples. His nephew, now known as Pliny the Younger, recorded the last day of an extraordinary scholar who died in the act of collecting facts up to the last minute of his life.

Pliny the Elder just couldn't keep away from the erupting volcano, such was its scientific interest to him. He ordered a boat to be made ready and invited his nephew to join him. 'Ashes were already falling, hotter and thicker as the boat drew near, followed by bits of pumice and blackened stones, charred and cracked by the flames,' writes Pliny the Younger. Pliny the Elder ordered the boat to hug the shore so he could observe the volcano closely. 'The flames and smell of sulphur, which always tells you the flames are coming, made the others run away ... When daylight returned on the 26th – two days after the last day he had been seen – his body was found intact and uninjured, still fully clothed and looking more like sleep than death.' He was the first person to die in the cause of scientific investigation.

Cosmo, your search for the truth about witchcraft is admirable, but do not risk your life or your bright future in the cause of your investigations. Take note of what Pliny the Elder once said, 'The best plan is to profit by the folly of others.'

Cosmo's face turned red. 'Tell me Jane, how a spirit could eavesdrop on our conversations. You're a mistress of eavesdropping, and a pretty one. After all these yards we have done together you're hiding something from me.'

I just looked at my feet.

'Lost your voice, again? Salus has learned to read mirror writing, why can't you?'

I was surprised to learn that Salus was taking interest in the scrolls.

'All I'm trying to do is to shield an innocent old woman from a volcano kicked off by some rascal girls. Some people would definitely benefit by this folly. It has now its own momentum and no one can stop it.'

'Aeee.'

'Turned into Jin, have you? Get out of you supernatural delusion and come back to the real world. Spirits are not going to solve the problems of this world. Only humans can do it, but only when they shed their erroneous beliefs.'

He started reading again.

A motionless earth at the centre of the universe, lost in cycles and epicycles, in a tumultuous, religious world sinking into the Dark Ages.

The stationary earth is at the centre of the universe. This erroneous belief dominated our thinking until the sixteenth century. But why?

'The earth does not rotate; otherwise objects will fling off its surface like mud from a spinning wheel. It remains at centre of things because this is its natural place – it has no tendency to go either one way or the other. Around it and in successively larger spheres revolve the moon, Mercury, Venus, the sun, Mars, Jupiter and Saturn, all of them deriving their motion from the immense and outermost spheres of fixed stars.' So proclaimed Claudius Ptolemy (*c.* AD 90–170), the last of the great Greek scientists. In his book, usually called by its Arabic name, the *Almagest* ('the greatest'), he synthesised the work of earlier Greek astronomers, and enhanced this synthesis with the results of his own observations and ideas. The *Almagest* was the standard textbook of astronomy for fourteen centuries. Its eminence, importance and influence can only be compared with Euclid's *Elements*.

A major part of *Almagest* deals with the mathematics of motions of planets. Ptolemy explained the wanderings of the planets by a complicated system in which the planet moved in a small circle called the epicycle, which in turn moved in a large circle around the earth. These cycles and epicycles harassed astronomers for centuries.

'What're cycles and epicycles, Cosmo?' I asked.

He took my hand and spun me around him as we moved around a garden gnome. 'An epicycle is a small circle that moves around a larger circle. We're in a cycle, but you're also in an epicycle.'

'So I'm like Ptolemy's Venus going around in a small circle which is going around the earth in a larger circle.'

'You move in spurious circles, my Venus. The true Venus goes around the sun in an ellipse as Thales will tell you when he talks about Kepler.'

'How do you know what Thales will talk about?'

Although Ptolemy's complex system of the motions of plants was wrong, it was a sophisticated mathematical model that fitted observational data collected by him and earlier Greek astronomers. As any true scientist would do, he made careful observations and compared them with those of others. Undoubtedly, in the *Almagest* Ptolemy was his scientific best. Yet he arrived at a wrong hypothesis.

'Chara, imagine yourself standing on the earth at the centre of the heavens. What do you see?'

'All the stars of the half of the celestial sphere above my head.'

'Can you see the other half which is below your horizon?'

'No.'

'As the earth turns round, the stars over your head will change. New stars will come into your view and others will disappear. What if you were standing at either of the poles?'

'I won't see any change in the stars.'

'While I'm standing outside my house I see things differently. I say some stars are rising and some stars are setting.'

We have two different systems, each of which would explain all the facts of the motions of heavenly bodies observed by Chara and me. One of the systems requires that the celestial sphere turns uniformly around an invisible axis. The other system requires that the celestial sphere remains stationary while the earth at the centre rotates about the same invisible axis.

Ptolemy knew that either of these systems could be explained by the observed facts. Yet the astute mathematician and astronomer could not detach himself from the great Aristotle's (wrong) idea that heavy objects fall towards the centre of cosmos faster than the light ones. If the earth were not at the centre of the universe, argued Ptolemy, it would fall towards the centre. Not only this, the massive earth would fall faster than things on it, leaving everything floating in space. Therefore, he placed the earth at the centre of the universe, where it remained until Copernicus displaced it.

After Ptolemy, Europe had started sinking into the Dark Ages; and the fast evolving Catholic Church wholeheartedly accepted this view. For fourteen centuries no one dared to challenge the Church's doctrine: we humans and the planet we inhabit are at the centre of the God's creation. With great zeal the Church would use this doctrine to silence science, again and again.

The doctrine was challenged when Copernicus asked himself a simple question: Could the observations of Ptolemy be explained by a system in which the sun, rather than the earth, is at the centre of the universe? Copernicus' challenge earned him the displeasure of the powerful Church

Ptolemy, however, supported Pythagoras' view that the earth is spherical. He gave many reasons for supporting this view, but the most ingenious was that if the earth were flat sunset and sunrise must take place at the same times, no matter in what country you were. He proved that the times of sunset and sunrise changed greatly as the longitude changed.

He also wrote a treatise on reflection and refraction. Now lost, *Optics* was the first attempt at the formulation of the laws of refraction. He can also be hailed as the greatest geographer of the antiquity. He drew projection maps which replaced plane maps used by ancient geographers. When Columbus saw the map of the known world drawn by Ptolemy, he was inspired to travel to the East by sailing west. From *Almagest* he knew that the earth was round; he hoped that its circumference were small enough to allow him to reach India before his ships would run out of food and drinking water. Columbus' estimate of the circumference was based on Ptolemy's erroneous estimate. Instead Columbus reached America, calling its aboriginal population 'Indians'.

~~~

On Tuesday morning I was again in Ware in the company of usual suspects, Vicar, Mrs Gardiner and Mrs Bowes – and a dodgy man of law. Mr Bowes had been helping Mr Pollard in witnessing the statements made by many town folks and believed, erroneously as Thales would say, that he had now enough evidence to show his authority as the parish constable. Vicar had asked Salus to come along but he refused on the pretext of being too busy in the clinic; Mama as always was keen but Salus would not allow her. Poor Cosmo never had any voice in this matter; diffident apprentices' utterances are always drowned in their masters' cacophonies.

The second time around Mrs Marsh was prepared for the invasion of her house and her privacy by this troop of trolls. Once the trolls were perched on chairs in her parlour, she didn't, as she had done in the past, brought them large mugs of ale or beer, but stared at Vicar. This made Vicar uncomfortable. He fumbled through his Bible and then asked, 'Mrs Marsh, we would be obliged if you could recite the Lord's Prayer.'

She nervously started reciting, 'Our Father which art in heaven, Hallowed be thy Name ...' and stopped after 'Forgive us our trespasses'. 'I cannot go further because you are trespassing on my hospitality,' she said with a touch of irritation in her voice.

'We're sorry Mrs Marsh to put your through this test,' Vicar's replied in a soft voice, which I thought his vocal chords were not even capable of producing. 'If you recite the Lord's Prayer accurately and without stumbling, I'll ask the constable not to take the depositions he has collected to Sir Henry Gilston. Did you know as a justice of the peace he has the power to issues a warrant for your arrest?'

This made Mrs Marsh more nervous. Trembling with terror, she started reciting from the beginning but stumbled over the words 'deliver us from evil' again and again and failed to complete the whole prayer. It seemed the thought of prison had made her incapable of thinking or speaking clearly. It became obvious to us all that she has flunked this important test for witchcraft Vicar had sat her. Failure to complete this holy task was a sign that she had an association with the Devil.

Vicar left the house without saying a word. Once we were in the coach and on our way to Hertford, Vicar asked, 'Bowes, how many depositions have you collected so far?'

'Five, Your Grace.' Mr Bowes always addressed Vicar as if he were the Archbishop of Canterbury. My foot! A vicar was a parish priest.

'Collect a few more and take them to Sir Henry as soon as possible. Come back here tomorrow with Jane, Mrs Gardiner and Mrs Digby to scratch the witch. '

A church parish was a unit of local government. Our plodding constables were at the lowest rung of parish office-bearers. We saw policing as un-English; in France it was a tool of royal tyranny. Yet our parish officers were not immune from using their constables as a tool of bullying the poor parishioners. By waving his red-coated wand Vicar had magically placed an innocent woman at the mouth of a dark dungeon.

Vicar looked at Mrs Bowes. 'You shouldn't come with your husband tomorrow.' Her bright face suddenly turned dark. He then muttered to himself, 'The girls don't want to see a man of God. I don't blame them. They are not themselves. God will forgive them. Amen!'

'Amen!' we all repeated.

~~~

In the evening when I walked into the consulting room I saw Cosmo with a new scroll in his hand talking to Salus and Mama. He gave me an enigmatic smile and started reading it.

A brilliant and charismatic scholar stripped naked, hacked to death with broken bits of pottery, her corpse torn to pieces and her remains burned.

Today's story about the first notable woman in science and mathematics is uplifting as well as unsettling. It will inspire Jane, sorry, all girls, who like your humble narrator Thales, dream of dabbling in science and mathematics.

Her name was Hypatia (c. AD 370–415). She was the daughter of Theon, a distinguished mathematician and astronomer at Alexandria's Great Library. Previously a Greek colony in Egypt, Alexandria was now a part of the Roman Empire and a turbulent mixture of cultures. Christians were in a majority. There were also Jews, Agnostics and the persons whom the Christians regarded as 'pagans'. Most of the pagans were either believers in the ancient Greek gods or were followers of various schools of Neo-Platonic religious philosophy. Neo-Platonism, based on the teachings of Plato and later Platonists, was born in the second century AD.

Very little is known about Hypatia's early life. In her thirties she became the leader of Neo-Platonic School, a close-knit group of intellectuals and aristocrats. She was a charismatic teacher and lecturer. Every evening she would ride in her chariot to her lecture hall and would transfix a large crowd of students who had come from as far as Rome and Athens. They were there to listen to her discussing ages-old philosophical questions such as 'What am I?', 'Where am I?' and 'What can I know?' which have not yet been answered.

The Neo-Platonists believed in a supreme power, the Absolute, which was mystic, remote and unapproachable directly by human beings. Hence there existed between the Absolute and the humans lesser gods. The first was Nous or Thought, which was an image of the Absolute. Next were the triad of Souls, which pervaded the material universe. Matter was considered evil, while the triad of Souls was pure. Humans were a mixture of the material and the spiritual. By self-discipline and control of the sense, humans could receive from the Absolute the secret of divine reality and free themselves from the slavery of matter. None of Hypatia's philosophical writings have survived and we do not know her views on Neo-Platonism.

She was also an outstanding mathematician. Only the titles of her mathematical works have survived, but sources describe her as a mathematician who eclipsed her famous father's talents. She wrote commentaries on the three greatest mathematical works of the ancient world: *Arithmetica* of Diophantus, the *Conics* of Apollonius and the *Almagest* of Ptolemy. She invented, among other things, a plane astrolabe to measure the position of stars, planets and the sun.

Hypatia was a pagan in an increasingly Christian city. Cyril, the new power-hungry Archbishop of Alexandria, saw her popularity as a great threat to his hold upon the citizens of the city. He and his followers also resented the fact that a woman was lecturing. They spread the rumours that she was a witch who practised black magic and cast Satanic spells on the people of Alexandria.

One dark night in 415, on the way home from the Library, she was dragged off her chariot by a mob of Christians zealots, stripped naked, hacked to death with broken bits of pottery, her corpse torn to pieces and her remains burned with straws.

Many works have perpetuated the legend that Hypatia was not only intellectual but also beautiful. Legend also has it that she was determinedly celibate, even repelling one suitor by confronting him with one of her used sanitary napkins as symbol of female body's physicality.

Hypatia's death marked the end of the golden age of Greek science. It also marked the end of the Great Library. In 641 the Arab armies of the Caliph of Baghdad not only razed the Library buildings, but burned the books to heat the public baths. None of the scrolls survived.

Europe had now entered into a long period of ignorance and inaction which is justly known as the Dark Ages which would last a thousand years.

Cosmo mouthed something inaudibly which only Jane the lip reader could understand as 'Herfordshire has now entered the Darkest Age'.

Salus was all smiles but Mama didn't seem happy. 'I do not believe in women going to universities and lecturing. A woman's place is in the home.'

Salus wisecracked. 'And the church.'

'That reminds me that Vicar Tayler has asked me to go to Ware to witness the scratching test to the witch. John, I'm going whatever you say.'

'You're a free spirit, gentlewoman. Has anyone solved the mystery of the spirit that calls itself Thales?'

Blood

'Do you believe honestly, love, I'm a witch?' She looked at me, her eyes filled with despair. 'They say the young ones are not as superstitious as us oldies.' Her momentary confidence waned and she started sobbing. 'I never did any harm to anyone.' Her sobs of despair turned into wails of a waif. Marooned on an island in sea of sadness, a storm of sorrow had blown away everything, leaving only bitterness at the bottom of her lonely life.

'Nnnn … nnnn,' I stuttered. 'Believe me, Mrs Marsh, I swear by all I hold sacred that you're not a witch.'

She composed herself and asked in a detached voice, 'Why are you here then, Miss Jane?'

'Mr Bowes has asked me to take you to Mr Bartlett's house.'

'Why?'

I was lost for words. I didn't have the courage to describe the ordeal she would go through at Mr Bartlett's house: five girls scratching her face or arms with their fingernails until blood was drawn. I didn't even know whether the practice was lawful though I was acting at the behest of our constable.

'You don't have to tell me nothing,' she looked at me as if I was a contemptible creature. 'Your friend James was here last night. He has told me everything.'

I was shocked to hear Cosmo's six-mile nocturnal walk. The little demon was acting as a free spirit, not even bothering to tell his best friend. Here I was working as a servile servant of the law. Law? Yuck! Whose side was I on? In my heart I wanted to help her. I was clueless: was it better to be on the inside looking out or on the outside looking in?

'A sly, selfish girl like you doesn't deserve an intelligent and decent lad like him.' she continued. A short introduction and my Cosmo had become a paragon of virtue in her eyes. A true sorcerer!

'I had asked her to see you,' I lied like a conniving bitch, my true image in her eyes. I had to do something drastic to redeem myself. 'We both want to help you. Really we want to help you.'

'How?'

'James has a plan. Didn't he tell you?'

'He said nothing about a plan.'

'He's going to talk to the girls and warm them to stop accusing you.'

'Will it work?'

'Yes, of course, it will work. You have not seen his power over girls. I'm one example right in front of you. My Cosmo is a sorcerer when it comes to girls, can bewitch them faster than anyone, even you, Mrs Marsh.'

A noticed a faint smile in her eyes. My magic was working, the magic of lies.

'I have no recourse, have I?'

'You can always say no.'

'How could a poor, old widow fight against powerful people of the town?'

'We're not a nation of savages; we cherish our laws.' Though the words gushed out of my mouth I was ashamed of saying them. What was about to follow at Mr Bartlett's house was nothing but savagery, savagery based on the catty comments of some girls. If Cosmo the scholar would have heard me, he would sneer, 'What chance a wretched woman has in our courts of law that gave privilege to property before human life? Did you know that a child could be hanged for stealing a handkerchief worth one shilling?' His tone would become sarcastic. 'You can't say our courts discriminate against children. They value the testimony of children.'

There was no sarcasm in Mrs Marsh's voice, only pity at my blind belief. 'You speak like a doctor's daughter, love. Sadly our laws do not cherish the destitute.' She hesitated for a few long moments. 'I might as well as go with you. James said that I must go along with them otherwise it would complicate the matters. Why do you call him Cosmo?'

'Haven't you noticed the effect of his cosmic powers over you?'

She gave me a cute toothless smile.

~~~

When Mrs Marsh and I entered Mr Bartlett's sitting room it was already crowded: two sets of bewildered parents with their five bewitched daughters (not to mention five spiteful spirits hidden in their bodies – or minds) and two anxious witnesses, Mama and Mrs Gardiner. Mr Bowes, representing the parish authorities, was standing near the fireplace with a large pewter tankard in his hand and talking to a maid who seemed exasperated. Perhaps he was griping about the slow service, not realising that he was a guest in a house, not a brash drinker in an alehouse. The Bull's best ale – I'm sure Mr Bartlett must have reminded him about the quality of the ale he served to his honoured guests – was free on that day and he was not going to stand there with an empty, or even a half-full, tankard. Typical behaviour of lowly officers when they found themselves in charge of an event.

Mrs Marsh's arrival signalled the rising of curtain for the girls (and their spirits). They were perfectly calm before she had entered. But their reaction upon seeing her was instantaneous and tumultuous. They jumped up and down around her in a circle, chanting, 'Admit you're a witch, admit you're a witch, admit you're a witch.' A cat came from nowhere – it was Mrs Marsh's cat and I had not seen her following us – jumped up and sat on Mrs Marsh's shoulder. They girls now started chanting, 'Save us from Satan's servant, save us from Satan's servant, save us from Satan's servant.'

The word Satan woke Mr Bowes up from his drunken stupor. He grabbed his baton and waved it in the air and shouted like a madman, 'Shut up girls, if you don't I'll throw you all in the gaol in the old castle. Rats bigger than rabbits will chew you in minutes.' They girls immediately became silent. No one, girls or spirits, fancy the prospect of being eaten by monstrous rats.

Mr Bartlett gazed at PC Plonk's antics in admiration, just short of standing up and shouting 'Encore! Encore!'

Pleased with the result of his policing prowess, Mr Bowes now bellowed, 'Amy, go to your room with your sisters and friends and wait there till I call them one by one.' The girls meekly followed Amy out of the hall. He then turned towards Mama. 'When you're ready ma'am, I'll bring in Alice.' Mama looked around like a presiding judge. 'We're ready, Mr Bowes. We can't delay much longer. Vicar wants to hear from us tonight.'

Mr Bowes went out and came back with Alice. The little Alice was subdued, still petrified by rats. She had been blindfolded so that she couldn't see who she was scratching. When she stood in front of Mrs Marsh, who was sitting on a chair, she turned into a feral rat. She gnashed her teeth in anger and scratched Mrs Marsh's arm.

Mr Pollard, who seemed to be enjoying the spectacle, rolled up his sleeve and came near Alice and placed his hand to Alice's hand. Alice wouldn't touch his hand. Encouraged Mrs Gardiner also tried, but Alice didn't touch her hand. Annoyed by these tricks, Alice fumbled about Mrs Marsh and when her hands and scratched Mrs Marsh's face so violently that her long nails broke down into splinters. Mrs Marsh's face was covered in blood.

Mrs Marsh was cooperating in her scratching. She even stroked Alice's hand and said, 'Take more blood, child. God help you.' Ungracious Alice hurled back, 'Pray for yourself, witch. Your prayer can do me no good.'

I had to witness these bloody scenes four more times. Please spare me from describing them again.

~~~

After a few days we heard from Mrs Gardiner that the news was bleak. Scratching had not cured the girls, as it was expected to do. They were still suffering from seizures, which were becoming more frequent day by day.

'I still don't know the medical reason for their seizures,' Salus sighed.

'You shouldn't worry, Dr Digby,' Mrs Gardiner consoled him. 'I pray to you, sir, please accept the fact that the girls have been bewitched.'

'Leave it to Vicar,' Mama consoled him. 'He is your old friend. He has promised me that he won't send Mr Bowes to the justice of the peace until you agree with the depositions.'

I was thrilled. 'Salus, please, please, tell Mr Bowes to burn them.'

'Jane, the situation is more complex than you can imagine.' He looked at Cosmo who was going through drawers. 'What are you looking for, Cosmo?'

'The magnifying glass you bought in Holland from that famous spectacle maker, sir.'

Salus took out a magnifying glass from his drawer. 'Here it is, Cosmo. Did you know why magnifying glasses are called lenses?'

'Because they are shaped like the seeds of a lentil.'

'Right. Why were you looking for it?'

Cosmo took out a scroll from a drawer. 'In this scroll Thales talks about lenses. I thought I will use a lens to magnify the writing to find out who has been writing these scrolls. It has to be someone in the town.'

Mama clapped her hands. 'Clever lad! Why do you think it is not a spirit, love?'

Cosmo deflected the question. 'Thales talks about his spirit now living in a toad. I have been wondering how a toad could write like us.'

Mama smiled. 'A witch's toad can do anything. Be careful, love. Come on Mrs Gardiner, we will enjoy coffee while these swots talk about science.'

I had this outrageous suspicion that Mama liked Cosmo more than her daughter. A jealous bitch I was.

Not so dark ages. While science stagnates in Europe, it flourishes in a world which doesn't demand science's obedience to religious dogma.

While the Christian Church assumed itself the depository and arbiter of knowledge and resorted to civil power to compel obedience to its doctrine, religious beliefs in China, India and the Islamic world did not become stumbling blocks in the intellectual advancement. Science stagnated in Europe's so-called Dark Ages, but it flourished in these countries. My story is about science in the West, but Chara reminds me, it would be rude of me not to mention advances in the East. Because of lack of time, I'll talk briefly about four discoveries.

While dark clouds of superstition had descended on Greek knowledge in Europe, Arabic scholars became its custodians. I'll start in Egypt with the 'mad master' of optics whose work marked the beginning of the science of optics and influenced Western scientists for centuries.

Hassan ibn al-Haitham (965–1040; known in the West as Alhazen) was born in Basra, Iraq. According to a popular story, al-Hakim, the Caliph of Egypt, invited him to Egypt to build a dam to regulate the annual flooding of the Nile. When ibn al-Haitham realised that Nile was far too wide to build a dam, he pretended to be mad to escape from the wrath of the cruel and eccentric al-Hakim (known as the Mad Caliph, he once ordered the killing of all dogs because their barking annoyed him). After al-Hakim's death in 1021 ibn al-Haitham lived a normal life in a house in Al-Azhar University in Cairo, writing, experimenting and teaching.

In his greatest scientific work *Kitab-al-Manazir* (The Book of Optics), written in 1038, ibn al-Haitham rejected the theory of the ancient Greeks that vision is the result of the eye giving out rays of light which allow the eye to see. Instead, he proposed, correctly, that vision was made possible by rays of light reflecting from an object into the eye. He suggested that the physics of reflection and refraction of light needs to be consistent with the biology of eye. He also explained how lenses work and attributed their magnifying power to the curvature of their surfaces.

In 1044 an unknown Chinese scientist wrote a book *Wu Ching Tsung* which included a recipe for a black powder. This earliest form of gunpowder was a mixture of saltpetre (potassium nitrate), charcoal and sulphur. The Chinese filled bamboo tubes with the black powder and used them as explosive projectiles thrown by catapult at the invading Mongols. These proto-rockets didn't do much damage to the enemy except scaring their horses. Towards the end of the thirteenth century the Chinese invented metal cannons using them to fire projectiles from their barrels.

While the Chinese soldiers were polishing their cannons, an unknown Chinese mathematician was playing with rods and beads. One day he turned them into a deceptively simple calculating device, the abacus (*suanpan* in Chinese). Simply a rectangular frame of rods with beads strung on them, a typical Chinese abacus has thirteen columns, each one divided into an upper deck and a lower deck. The lower deck has five beads in each column; the lower deck, two beads per column. The beads in the upper deck are worth five times the beads in the lower deck. The first column represents single units, the second tens, the third hundreds, and so on. This means you can do calculations with numbers up to ten trillion. If you are an expert, you can carry out not only addition, subtraction, multiplication and division problems, but can also work out fractions, and square and cube roots.

My brief journey to the East ends in India in *c.* 1150, when Bhaskara II (1114–85; known as Bhaskaracharya, 'Bhaskara the Learned') proposed a perpetual motion machine, a machine that would run forever without consuming energy. Bhaskara has a respected place in the history of mathematics. He wrote the first works using the decimal number system which described rules for calculating with zero and the concept of negative and positive numbers.

He also has a special place in the annals of perpetual motion machines as the designer of a wheel that could turn forever, although he never built it. The wheel, with containers of mercury around its rim, was designed to rotate constantly, because the wheel would always be heavier on one side of the axle. This idea, like the ideas of zero and decimal numbers, re-appeared in Arabic writings. From the Islamic world it reached the Western world. Over the centuries numerous scientists and engineers tried – and failed miserably – to build perpetual motion machines. Perpetual motion defies the sacrosanct law of conservation of energy that no machine can produce more energy than it uses.

Chara laughs. 'Spirits like us do not have to obey any laws of science.'

'Hertford's Hypatia, did you know that there were great civilisations beyond our shores while we lived in intellectual darkness caused by our inability to see beyond our scriptures?' Cosmo remarked.

'We're still living in darkness, Hertford's Hermes. Renaissance might have revived arts and literature, it has not dimmed our belief in witchcraft and sorcery,' I replied.

The golden ratio, the dance of zero with nine Arabic numerals and a mathematician who shone brilliantly through Europe's Dark Ages.

Your Thales is back in Europe to introduce you to Leonardo Fibonacci (*c.* 1170–*c.* 1250). An Italian mathematician, he is now best known for the simple series of numbers which comes from the answer of a puzzle posed by him: 'Beginning with a single pair of rabbits, if every month each productive pair bears a new pair, which becomes productive when they are one month old, how many pair of rabbits will there be in one year?'

The series (1, 1, 2, 3, 5, 8, 13, 21, 34, 55, 89, 144 ... add the last two to get the next) is known as the Fibonacci numbers. This sequence of numbers has many other interesting mathematical properties. For example, the ratio of successive terms (larger to smaller; 1/1, 2/1, 3/2, 5/3, 8/5 ...) approaches the number 1.61803398887... . This ratio is known as the golden ratio and is denoted by the Greek letter phi. Like pi, phi is an irrational number. The decimal sequence of an irrational number goes on forever and does not repeat in any permanent pattern.

The story of phi began with Euclid. He showed how to divide a straight line AB into two by a point C so that the ratio of the longer segment (AC) to the shorter one (CB) was exactly the ratio of the entire line (AB) to the longer segment (AC). Irrespective of the length of the line, the ratio is always equal to phi.

Ancient Greeks used phi extensively in art and architecture as they considered it pleasing to the eye. The exterior dimensions of the Parthenon in Athens, built in about 440 BC, form a perfect golden rectangle (a rectangle whose sides are in golden ratio).

Curiously, phi also appears in the natural world. Flowers often have a Fibonacci number of petals (look at the arrangement of florets on a cauliflower). The seeds on a sunflower are arranged in two sets of spirals. The ratio of the number of seeds in the two spirals is phi – and so is the ratio of your height to the distance from your belly button to the tip of your feet. When a falcon dives towards its prey, it swoops in along a path that is mathematically related to phi.

During his travels in North Africa Fibonacci learned of the decimal system that had been evolved in India and had been taken up by the Arabs. Upon his return to his home town Pisa he published a book *Liber Abaci* (1202) in which he introduced to Europe the Arabic numerals we use today. 'These are the nine figures of the Indians: 9 8 7 6 5 4 3 2 1. With these nine figures, and with this sign 0 which in Arabic is called *zephirum*, any number can be written, as will be demonstrated,' he wrote in his book.

Except that he lived in Pisa, it's a great pity that we do not know more about this remarkable man who even in Europe's Dark Ages formed a link between the East and the West. Luckily we know about his mathematics from his five books which have survived.

'Why are you not trying your lens on cauliflowers, sunflowers, mollusc shells and belly buttons to discover the golden ratio in nature?' I asked Cosmo.

'Thales didn't talk about mollusc,' Salus remarked as he walked out of the room.

'Knowledge is not the preserve of the privileged few like you, Salus,' I grizzled.

Salus poked his head through the door. 'It was until Gutenberg came along with his printing press. O Thales I pray thee, tell this grizzling girl how Gutenberg liberated knowledge from the preserve of the privileged few.'

When Salus was out of sight, Cosmo stood up and looked through his lens. 'Tell me, my cry baby, when I'm going to meet Amy. Your lies will magnify thousand times through this lens. Be honest.'

'Water Row in Ware on Christmas Eve.'

'Why Water Row?'

'Every year they have a fair there. Surely we will find the girls at the fair or wassailing in the streets.'

'That's two weeks away. I'm worried Sir Henry Gilston would have issued an arrest warrant by then.'

'Nope. I overheard Vicar talking to Salus. Vicar is off to Cambridge for a few weeks and he wants to be around when the constable presents depositions to Sir Henry Gilston. So nothing will happen before Christmas.'

'Your ill-gotten knowledge has some good news.'

'And the better news is that Mama is going to London to be with her sister at Christmas. No one to clip my wings, I'll be a free bird. I could go with you without any drama.'

I doubt Cosmo heard me. He was busy examining the scroll with his lens.

Bells

On the morning of the fair Cosmo and I were dressed to kill; colourful bands in Cosmo's beaverskin hat and feathers in my woollen felt hat crowned our attire (snazzy for our small town, may be not for the big smoke). When Salus found out that we were walking to the fair, he insisted that we take his coach as he didn't like the idea of his sole heir walking through the woods and fields on a wintery morning. 'Mr Bowes is on duty at the fair, Jane. I'll ask him to keep an eye on you. Fair crowds get rowdy sometimes.'

I shuddered at the thought of a drunken Mr Bowes breathing all over me and poking his nose, the red button that wallowed in fanciful perfumes of buxom beauties of his bawdy books, in my affairs. 'No, no, Salus.' I winked at Cosmo. 'Your precious princess can do without the king's guards when she has a gentleman and future physician at her side.'

Unlike most parents who pretend to have no memory of their youth, my father understood the ways of the young people; Mama would have never let me go with Cosmo without a chaperone. 'Very well, I won't ask him. Look for him if you need help. Won't you, Cosmo?' Cosmo nodded in earnest.

A coach stopped with an awful din in front of our house; thinking it was our coach we all stepped out. A young maid came running out of the coach. I recognised her as one of Mr Bartlett's maids. Breathless, she had to wait a few moments before words came out of her mouth, 'Dr Digby, sir, Mr Bartlett wants you to come to the Bull right now.'

Salus patted her consolingly on her back. 'Calm down, child. What has happened?'

'Mr Donne is terribly sick, sir.'

'Donne who?'

'Sorry sir, Mr Donne is the brewer. Since he started unwitching the Bull's ales and beers after the incident at the Crown, he has not been feeling all right. This morning after delivering a beer barrel he fell down on the floor.'

'Delivering beer in the wee hours of the morning?'

'Mr Bartlett has been stocking up for the fair, sir.'

'What happened after the fall?'

'He felt dizzy for a while and when he got up he said he heard a woman's voice say, Donne, you have been plunging red-hot irons into my body through your barrels, now it's my turn to plunge those irons back into your body. Mr Bartlett thinks he has been cursed by the witch. You know who I mean.'

'How is Mr Donne now?'

'He's still in pain and lying on a bench at the Bull, all covered in blankets. Mr Bartlett wants you to come right now. He has sent this coach for you.'

'Is the pain in his chest?'

'I think so, sir.'

Salus looked at Cosmo. 'What's your diagnosis, gentleman doctor?'

'Delusion caused by headache if his head hit the floor, or may be a minor heart attack causing momentary blackout, sir. The bit about the witch is bull. '

Salus laughed, an innocent and infectious laugh, unpatented like his orangeypills but guaranteed to make you feel happy. 'You're probably right, Dr Cosmo.' He gave me a peck on cheeks and pointed to our coach which had been waiting behind Mr Bartlett's coach. 'Jump on. I wouldn't want to hold you any longer.'

'Sir, I can cancel my program, if you want me to go with you,' Cosmo said in his most sincere voice.

'I may be your master, young man, but I wouldn't dare to ask you to work on your day off. Off you go. For me, merry Christmas means more patients in my consulting room.'

Once we were inside the coach Cosmo said to the coachman, 'Reg, first to the church.' He then he turned his worried face towards me. 'We won't be long. We have to say a short prayer.'

I goggled at him. 'Prayer?'

'Prayer for Mr Donne. If he has had a heart attack, it is possible he could have a massive heart attack within days.'

'So what?'

'If he dies, everyone will blame Mrs Marsh for his death. She will be charged with his murder by witchcraft.'

'Just like that.'

'Yes, just like that. Instead of you burying your head in science and medical books, I wish you read books about witchcraft trials. There're some around.'

'But how will our prayer help Mr Donne?'

'Don't you believe in the power of prayer, I mean, distant prayer, you ignorant heathen? We have to do our best to make sure that Mr Donne lives; otherwise this excursion to influence the girls would be a waste.'

I never knew that my gentleman who mocked things supernatural was so deeply spiritual. At the same time his sarcastic comment perplexed me (and still does). I understand, like Salus' orangeypills, if you're sick your faith could help you heal as believing in a cure leads to real changes in the body. But how could strangers' prayers help heal a sick person miles away, especially when the sick person doesn't even know that they are being prayed for? If it were possible then the opposite would also be within the realms of possibility: the evil prayers of a person, or a witch for that matter, could also harm someone miles away.

Anyway, I closed my eyes and started reciting silently the Lord's Prayer, not for Mr Donne's heart, but my heart-throb's heart. There was one consolation at least: as I could recall the prayer verbatim, I couldn't be accused of a being a bitch, er, witch.

~ ~ ~

Outside the fair entrance a signpost said, 'Lords, gentlemen and loons. You're welcome to our town'. I pointed the sign to Cosmo and giggled, 'Women aren't welcome here. Let's go back.'

'But loony girls are.' He took my hand and dragged me towards a tent outside of which a portly man with enormous handlebar moustache was spruiking the singing skills of conjoined twin sisters captured in the jungles of Malaya. Conjoined twins and other 'freaks' such as the excessively short or tall, people whose gender was not clear, contortionists, bearded ladies and tattooed men were all standard part of our fairs, circuses and even zoos. As my medical curiosities did not extend to unusual bodies which were considered symbols of supernatural creation, I continued walking leaving Cosmo behind to advance his medical knowledge of so-called monstrosities.

It didn't mean that I was not interested in human bodies, beautiful human bodies, though we English were renowned for our prudishness. Unlike other European cities, a randy tourist would find no nude statues of women in London, but he could always choose from thousands of whores who paraded their wares openly. Since I had heard about our PC Porno's collection of saucy books, I was keen to peruse one. If Salus had a secret trove of such books I had failed miserably to discover it.

Now it was my opportunity to discreetly acquire one from a chapbook seller. I walked through throngs of laughing lads and lasses past pedlars' booths with long silken laces hanging upon twine and tables of glittering toys, knives, combs, pins, scissors, gloves, amber bracelets and spectacles to read; past lotteries booths where silver spoons were won and rings of gold; and then past swings and roundabouts. As my swing and roundabout days were behind me I walked towards food booths ('Hot peascods', one began to cry) and bought a baked pear ('Smoking hot, piping hot/ Who knows what I've got/ In my pot?/ Hot baked wardens/ All hot! All hot! All hot!'). Nibbling it I stood in front of a quack peddling an elixir ('Found by the Providence of the Almighty this health-bringing drink has cured many credible persons ...'). One incredulous person refused to believe that his elixir guaranteed to cure gout, kidney stones, colic, shortness of breath, scurvy, coughs, wheezing, malaria, epilepsy, fits and numerous other illnesses while Salus' cupboards full of medicines had failed to cure five bewitched girls. I laughed and walked away when he diagnosed my illness just by looking at me, and tried to push a bottle in my hand with the spiel, 'Guaranteed to raise languishing nature and melancholy drooping spirits, miss.' Perhaps he was right. A lonely teenage girl stuffing herself with food had to be depressed.

Chapbooks, twopenny or threepenny little booklets, were highly popular in my days. Their subjects ranged from political and religious propaganda, astrology, palmistry, dream interpretation, cookery hints, beauty tips, courtship advice, plays, songs, fantasy and sensation to sex. I picked up a book from the back of the table and found it was a light-hearted ballad about a wine cooper's courtship of a merchant's widow. I blushed when I read:

He told her in his breeches,
There was a best of riches,
Right pleasant to behold.

I immediately put the book down when I saw Mr Bowes and another constable approaching. Wearing red coats and carrying black batons, they were patrolling the fairground crying out lustily, 'Look about you there', a warning about pickpockets and various cheats who were attracted to fairs like bees to honey.

When they had walked past without noticing me, I left the stall ignoring the stallholder's lecherous gaze and comment, 'I've more books like this in the box under the table for lovely lasses like you, miss.' Cosmo was nowhere to be seen so I decided to look for him.

I didn't have to go far. I saw him holding Amy's hand in front of a candy stall and the other girls were swarmed around candy jars. My mind was gripped by jealousy, the tyrant of the mind. To get closer to them, I pushed through a rowdy group of young men and women singing and dancing. A woman was singing:

If friends do frown and fret
And parents be angry,
And brothers' grief is great
Yet I love none but thee.

Playing a lute, a man replied:

What and if I be, boy,
I'm ne'er the worse;
She keeps me like a gentleman
With money in my purse.

I was all for romantic love, but that was not the time for warm feelings, especially when another woman was sizzling with my own love. I was relieved when I reached closer and heard Amy saying loudly, 'Let me go, sir, I don't need your help. ' The next thing I saw Amy kicking him in the shin, Cosmo muttering in pain, 'Amy, please listen to me. I'm a doctor, I can help you.' Amy kicked him again saying. 'You're not a doctor and will never be one, never. You stink like your vicar and the place for stinking vicars is up there.' She pointed towards sky and kicked Cosmo so hard that he was lifted up thirty feet – yes, as high as the roof of our house, I'm not exaggerating – and then hit the ground with a thump. I ran frantically looking for Mr Bowes.

When I came back with Mr Bowes, Cosmo was still on the ground but luckily he was not unconscious. Amy and her charges were nowhere to be seen. Mr Bowes lifted him up with ease – our nerdy constable had turned into a Hercules in my eyes – and carried him to the Bull.

Salus was still there; he examined Cosmo thoroughly and declared him fit except some minor bruises on his legs where Amy had kicked him. He laughed off my account of events and suggested that the height Cosmo fell on the ground was more like three feet. It made me worried about the state of my mind. Like the girls, has the wrench of the Witch of Ware wrecked my brain too?

If the witch hadn't controlled my mind yet, it was totally under her control when I saw Mr Donne, who had been sitting quietly on a stool and drinking copious amounts of gin, suddenly go into flame. Within seconds he was thoroughly incinerated. All that was left was a heap of ashes and his skull on the stool, and the blackened bones of his legs and four fingers on the floor. The room was filled with greasy, foul-smelling soot; everything else in the room was intact including the stool.

This awesome spectacle froze the noisy drinkers transforming the place into an oxymoron: a silent pub.

Mr Bartlett made the sign of the cross and whimpered, 'Dr Digby, sir, the witch's magic powers are out of control. You must help us, sir.'

Cosmo consoled me. 'I have had heard about spontaneous human combustion and I have now seen it with my own eyes. It's not witchcraft. Science can explain it.'

I was too scared to speak.

~~~

As our coach reached near Hertford in the evening, we were puzzled to hear the church bells ringing. Salus asked Reg to drive the coach to the church. Inside the church, Samuel Harsnet, Vicar's assistant, was surrounded by many men and women – worried faces, angry faces, subdued faces peering out of their dark winter clothes in the dim candlelight.

Mrs Bowes was telling everyone to draw or scratch a witch's knot on their doorways of their homes and stables. 'Make sure you draw the symbol in one continuous loop,' she exhorted the gathering. Another weapon in her armoury of protective charms, I thought.

Everyone hushed when they saw Salus. He asked Mr Harsnet why they had been ringing bells, but before he could reply Mrs Bowes came forward, curtsied and said, 'Didn't you know, sir, church bells drive witches away? They are so powerful witch-deterrent that the witches would fall off their broomsticks if they heard them ring.'

Salus seemed amused by this explanation. 'Amazing! What it has to do with Mr Donne's death?'

'Sir, I had asked Mr Donne's wife to burn the body of a sheep to undo the witch's spell.'

Malcolm Landish raised his hand. 'Permission to speak, sir. I was there when the sheep was thrown into the fire. Mr Donne's body incinerated exactly at the same time. No doubt that the witch burned poor Mr Donne. God bless his soul. The witch must be punished. Everyone here agrees.' Aye, ayes reverberated around the church.

Sensing her control over the situation, Mrs Bowes thundered. 'This witch is very powerful and crafty, sir. We need your help.' She then implored Mr Harsnet. 'Deacon, please beg Dr Digby to help us while Vicar is away.'

Mr Harsnet, a timid and shy man, looked at Mrs Bowes but when she saw fire in her eyes, he spoke even louder than his nemesis, 'Dr Digby, sir, on behalf of the town folks I beg you to send the parish constable to the justice of the peace to seek warrant for the arrest of the witch.' He then looked at the gathering and raised his hands, 'Do I have your support?' Again, aye, ayes reverberated around the church.

Salus looked uncomfortable. 'Sir Henry cannot issue arrest warrant without any evidence.'

George Gifford shouted angrily, his face turning blue, 'What evidence he wants now? Another innocent man going into flames?'

Mr Landish raised his hand again. 'Permission to speak, sir. I hear Mr Pollard has collected many depositions. We beg you to forward them to Sir Henry with your medical opinion.'

Mr Gifford took a few deep breaths, recollected himself and spoke in a measured voice. 'Dr Digby, you're a good doctor and everyone in these parts respects you greatly. You have pulled out many of us from the jaws of death. Pleased do not ignore our pleas. As a doctor it's your sacred duty to save us from this witch.'

Cosmo looked more and more despondent. He had probably surmised that the town folks now undoubtedly believed that Mr Donne had been murdered by a witch. His mission to save Mrs Marsh from the gallows seemed nearly as impossible as seeing witches flying on broomsticks. Cosmo and I had had spent many nights in the dark streets of our sleepy town looking for one. I didn't know of anyone who had actually seen a witch flying, yet almost everyone in my days believed in flying witches as much as they believed in God. But then again, no one had seen God. Our beliefs are not necessarily shaped by our senses.

~~~

Cosmo gave me a broad smile, his white teeth sparkling like lightening, when we arrived home. A scroll was hanging from the door knob. I was astonished to see how much he looked forward to the scrolls. Why was he seeking solace in science? Didn't he know that a sermon from the pulpit, however irrational, had more power over people than a well-reasoned scientific idea?

Keep it simple, stupid ... and if you contradict the scriptures, you may be ordered to appear before the papal court on charges of heresy.

How do you explain a scientific idea? Some will take you through a maze of facts, hypotheses and observations to arrive at their theories and you will still be doubtful. Others will cut the chase and direct you convincingly to their theories.

In the so-called Dark Ages of Europe, William of Ockham (c. 1290–1349), a Franciscan monk, developed a bright rule which is of vital importance in the philosophy of science even today. The rule – it is vain to do with more what can be done with less – implies that the number of causes or explanations needed to account for the behaviour of a phenomenon should be kept to a minimum.

It is a guiding principle in developing scientific ideas, and it insists that you should prefer the simplest explanation to fit the facts. The rule has been interpreted now to mean that when you have two competing theories that make exactly the same predictions, the one that is simpler is the better. In other words, the explanation requiring the fewest assumptions is most likely to be correct.

The advice – keep it simple, stupid – is in similar vein. But everything should be made as simple as possible, but not simpler. 'Trim fat, but leave flesh on the bones of your idea,' advises Chara.

Cosmo stopped reading and looked vacantly at me. 'It's vain to do with more what can be done with fewer. Useful advice.'

'What for?' I asked.

'If Mrs Marsh's case ever goes to the court, the idea that requires the fewest assumptions is likely to be correct.'

'What idea.'

'Miss Loon, the idea that has overpowered reasonable men and women.'

'Mr Bright Spark, I'm still in dark.'

'That Mrs Marsh is a witch.'

'It's not a scientific idea.'

'Never mind, we must prepare a deposition with fewest assumption for Sir Henry.'

He didn't wait for my reply and started reading again.

This razor-sharp rule is now known as Ockham's razor. William, a philosopher and theologian, came from Ockham, a village in Surrey, near London. In his youth he joined the Franciscan order and studied at Oxford, where he lectured from 1315 to 1319.

At Oxford, which was then a great Franciscan centre of learning, William became the leader of a school of philosophy called nominalism. Nominalism and realism are theories about the study of knowledge. Realism accepts universals or general ideas as they are. Nominalism, on the other hand, says that such ideas are mere names without corresponding reality.

St Thomas Aquinas, the thirteenth-century Italian philosopher and theologian who integrated the philosophy of Aristotle in the Christian theology, didn't go beyond the cosmology of Aristotle. Like Aristotle, he believed that nature abhors a vacuum. The cosmos was a sphere filled with matter throughout its volume, because any action on a body by a force required a direct or indirect physical contact. The first proof for the existence of God, he said, was that the motions of heavenly spheres required a Prime Mover. William rejected this view and said that a body in motion does not necessarily continuously require the physical contact of a mover. The action can be at a distance, like that of a magnet moving a piece of iron without touching it.

William also believed in world beyond our own planet, an idea rejected by Aristotle – and the church. He said that there could be no assurance that the world was finite, or that it had a governing unity, or that it was eternal, or that there were not several worlds.

'I wish I were in a world far away, without witches.' I thought but didn't say as Cosmo hated frivolous interruptions.

William's statements in his philosophical and theological writings, including the ideas on other worlds, aroused such opposition that he was refused his Master of Theology degree at

Oxford and was ordered to appear before the papal court on charges of heresy. In 1328 he fled to Germany and, according to a story, asked Emperor Louis IV for protection with the plea: 'Defend me with your sword, and I will defend you with my pen.' He remained in Germany for the rest of his life.

The printing press made books cheap and accessible, but the state and the church could still ban books. Will thoughts be ever truly free?

William's pen was potent, but its power was the power of one, not millions. In William's times books were copied by hand and only a handful of copies of each book was available. Most of the time these copies were not faitfull reproduction as many of the scribes fancied themselves as editors and made changes. Owning books was the symbol of wealth. The knowledge was the privilege of the few; the progress of new ideas snail-like.

Then came Johannes Gutenberg (*c.* 1395–1468) with his printing press, one of the most important technical breakthroughs in history. Within a short time books, which were sold in gold guineas, were sold in copper pennies, making chapbooks accessible to girls like Jane.

In Europe in the early fourteenth century there were only handwritten books or some short texts, printed using wooden blocks into which words had been carved. First time in history books could be reproduced without error.

Gutenberg was a young engraver and gem-cutter when he thought of using movable type to compose whole books. He experimented over several years, borrowing large sums of money to cover costs. He made moulds of letters for casting individual characters in metal. He invented devices for composing the types on a wooden plate and for inking the composition evenly, and finally a hand-printing press for making impressions of the plates on paper. He tested his press by printing an old German poem over and over again.

He was now ready for the job he'd been dreaming of for years: printing the whole Bible in Latin. He composed 1,283 pages of forty-two lines each, and printed 180 copies. The first book ever printed from movable type was ready in 1455. Proof that the first printed book was indeed a bestseller is found in a letter dated 12 March 1455, where the writer enthused: 'I did not see any complete Bibles, but I did see a certain number of five-page booklets of several of the books of the Bible, with very clear and proper lettering, and without any faults … even before the books were finished, there were customers ready to buy them.'

'I know there was immediate opposition to Gutenberg's remarkable invention.' Chara mutters.

The state and the church were fearful of the invention as they saw it as something that would undermine their authority. In England the king required all printers to seek approval before they could print anything. Up to this time the Bible was owned and interpreted by priests only. They were worried they would lose control over their parishioners. The easy availability of the Bible led to widespread criticism of the prevailing theological thoughts and influenced the development of schisms, which resulted in the Protestant Reformation led by the German monk Martin Luther.

However, printing presses were accepted readily and within years they were set all over Europe. Nobility and clergy were no longer the only repositories of knowledge. But it didn't stop them from banning books that challenged their beliefs.

'Eureka!' shouted Cosmo. I had never seen him so happy. 'You know what we should do?'

'Keep quiet, so we don't disturb Salus,' I replied.

'We should use Ockham's razor and Gutenberg's printing press to silence Sir Henry.'

'Tch, you mean we cut Sir Henry's throat with a razor and then pass his body through a printing press?'

'You have a wicked imagination. You should be accused of witchcraft, not a sweet old lady.'

'You're right and there is a victim right in front of me. For you I'm willing to trade place with her. Tell me what you have in mind. Keep it simple, stupid, before you go into flames. Your face is already turning red.'

'Your magic won't work on me. Listen carefully; what I have in mind is to print a pamphlet explaining in simple, sharp language that witchcraft is nothing but superstition and it's against the teachings of the Bible. On second thought, we would distribute it widely to the public.'

'Why?'

'To change the public opinion.'

'What happens if Vicar proclaims from the pulpit that your pamphlet is the work of Satan and orders the parishioners to go and piss on your pamphlet?'

'Forget him for a while.'

'Okay; there are no printing presses in Hertford, how would you print them?'

'I'll ask Mr Johnson. He is an enlightened man and likes talking about science. He goes to Oxford often.'

'Will he agree?'

'People who accept bribes are always amenable to ideas that bring in more bribes.'

'We don't have any gold coins to bribe him.'

'What happened to all those silver and gold trinkets you love buying?'

'I do not mind giving you a few trinkets, but I'm worried about Mama and Vicar. Salus will just laugh his head off at this silly idea.'

'We will do it secretly. No one will ever find out.'

'Who's going to write it?'

'We'll ask Thales.'

'Thales? Do you know him?'

'No, but we'll pray.'

'You think you can change the word simply by praying?'

'Prayers generate spiritual cosmic waves science will never find about.'

I put my hand forward. 'It's a deal, Mr Spiritual Power, if you answer one question: why did Amy say that you will never become a doctor? Did she see you in her crystal ball when she was divining her future husband's occupation?'

'I don't know. Why would she worry about my occupation?'

'She fancies you.'

'Didn't your Bible-loving Mama teach you that jealousy is cruel as grave?'

'I'm worried Amy might be right and you would never become a doctor.'

'Didn't Dr Digby teach you that it's worry not the hard work that kills people? Worry is like living in hell's fire. Mr Donne experienced it only for a few seconds. People who worry experience it all the time and combust slowly. Slow combustion is crueller than instantaneous combustion.'

We shook hands.

Knot

Prepare yourself for a long, monotonous monologue. No, no, not you, I meant Martha. She probably didn't hear me; the old age hadn't robbed of her acuity of hearing, it had only made it less receptive to voices except that of Mama's. My spunky Mama was back from London and she would be talking non-stop about her adventures and exploits in that glorified hive of three hundred thousand souls where slops are tossed from windows and darkest crimes are committed. (I tell you middle-aged married women do not have experiences that could be called adventures and exploits, unless they have been unfaithful to their husbands.) A monologue requires an audience (without an audience you would be called a loon and most welcomed at Ware). Besides cleaning and cooking Martha's other chore was to punctuate Mama's monologues with random interjections ('wow', 'really', 'unbelievable' or anything except 'encore' that made a boring account look, well, adventurous).

'Would you believe, Martha, I walked on the Thames?'

Mama walked on water! I feared our house would soon turn into a shrine to Saint Madame Digby.

'It was incredibly cold, so cold that the Thames froze and became a thick ice sheet for miles. Coaches were plying on the river as in the streets.'

Wow, I thought. If I had uttered this interjection I would have made Martha redundant.

'Then they came up with this incredible idea of a fair on the frozen river. Oooh, they had booths selling everything; and so many games, sliding with sleds and skates, horse and coach races, bull baiting, cock fighting, you name it and they had it. So much drinking and eating, it was a bacchanalian orgy. It was so much fun I shall never forget this carnival.'

I felt jealous of her. A bacchanalian orgy with Cosmo on the frozen River Lea; that was something to dream about.

'My sister's husband is a printer. He setup a printing booth and sold thousands of souvenir cards for six pence each. Expensive isn't? Though each card had buyer's name printed on it. You know Martha I was there when King Charles visited his booth with his family. He also bought a card.' She then showed a card with 'Mrs. Catherine Digby of Hertford. Printed by Wm. Croom at his Printing Press, at the Frost Fair on the Thames, on 3rd January 1684' printed on it in 'clear and proper lettering'.

Here was Mama with a real printed card while our idea of printing a pamphlet was not going anywhere. Why was the ghost of Gutenberg smiling on her?

'I saw a man digging a hole through the ice and putting his hand through it and catching a fish. Live fish, would you believe it? I thought the river would be frozen right up to the bottom.

They say that the ice was only a few inches thick, as much as eleven inches in some places. A miracle, indeed.'

Not a supernatural miracle, Saint Madame Digby, it's a natural wonder. Who could explain to her that practically every substance contracts as it becomes colder, but water expands to have about one-tenth more volume as a solid than as a liquid? If ice did not float, oceans and bodies of water would be frozen from the bottom up and there would be no living things in them. Without this amazing property of water, pardon the pun, a cold snap would snap off life from our planet.

Salus paralysed the monologue when he walked into the parlour. 'Martha, please go to Cosmo's house and find out why he is late. The lad has never been this late. I'm deeply concerned about his mother's health.'

~~~

Cosmo came in the afternoon, announced 'Nyx is dead' in a sad and subdued voice and slumped into a chair like a sack of spuds.

Salus seemed relieved to hear the news. Perhaps he was expecting the worst; since her husband's death Cosmo's mother had been visibly wasting away. 'Sorry to hear the sad news, Cosmo. Nyx was an affectionate dog, an obedient dog.'

'Thank you, sir. I gave him a decent burial. I carried his body in a push cart all the way to the river bank where I buried him under a very old chestnut tree.'

Mama looked at Cosmo in admiration. 'You did the right thing, love. How did Nyx die?'

'It's an amazing story, Mrs Digby,' Cosmo replied.

'It can't be more amazing than the stories I have heard from Deborah Bowes. You floating hundred feet above the ground and Mr Donne exploding in flames and leaving no trace behind. Did you know that the witch's second husband was also burned in hell like Mr Donne before he died? His ghost still haunts the Crown where he worked. Dorothy Sampson has seen him many nights in the cellar moaning and cursing his wife. They say her first husband drowned in the River Lea and his body was never found.'

Thy fibs, fictions, fantasies, Catherine. Devoted but bewitched parishioners of my town, Mama included, believed that the words of Bible were immutable, yet didn't care about the truth of the words they spoke. I forgave their liberties with words the as if they were the followers of the exotic Tantric philosophy which teaches that every word has four meanings: the literal meaning, the general meaning, the hidden meaning and the ultimate meaning. And you please forgive me for distracting you from Cosmo's story.

'A big thunderstorm woke me up very early this morning. I came down to make a cup of coffee. As I was putting logs in the fireplace, I was startled to see a blue-white ball coming out of the chimney. It was two feet in size and glowing like a candle. It must have squeezed itself through the chimney as our chimney is not that wide. After emerging from the fireplace it turned left, moved around the wall and then through the door into the kitchen. There it hit a bucket with a clang, went into the passage and then passed through the closed door. I tried to pick up the bucket but it was too hot.'

'Wow!' I said. It was a wow freshly baked with all the flavour and aroma of instantaneous exclamation, not one of Martha's musty wows.

Cosmo didn't acknowledge my hot wow; he had perfected the art of ignoring me in presence of my parents. 'I was scared but opened the door and looked outside. The ball was now speeding through the street and I could hear a hissing and crackling noise, Nyx barking madly and running after it. The next thing I remember Nyx running ahead of the ball, the ball hitting him and him lying dead.' Cosmo wiped tears from his eyes. 'His body was completely scorched and the ball had disappeared leaving a sharp, repugnant smell behind.'

'Was Nyx burnt like Mr Donne?' Salus asked.

He looked uncomfortable, but replied like a seasoned doctor, 'No, sir, it was a different kind of death. Mr Donne's body suddenly exploded as if a fire burnt inside his body. Nyx's body was like that of a cow that had been hit by lightening.'

Mama crossed herself and gibbered, 'Praise the Lord, the Devil didn't hit you. I'm afraid the witch is seeking revenge. You must have upset her somehow. You should draw a witch's knot on the fireplace to protect yourself. I'll ask Mrs Bowes for other ways to protect you from the witch.'

'Thank you.'

Salus' intense face betrayed his mind's turmoil. 'My mind is numb. I've never heard or read about such a ball. It seems more like a ball of spirit than a ball of matter.'

My mind was also in turmoil. The Devil, witches, demons, spirits, sprites, gnomes, elves ... they all now appeared to me too real to be ignored. My belief in science was shaking. Any kind of rational thinking or scientific reasoning as preached by Thales wouldn't have saved Cosmo if he had come into the path of the Devil's ball.

For Mama the road ahead was clear. 'John, Vicar will be back in a couple of days. You can't wait for him. Depositions collected by Mr Pollard and a word from you in their support will convince Sir Henry to put the witch in the gaol. Do it now. I mean now.' I had never seen such authority in the voice of my plucky Mama.

~~~

In a witch-crazed world an arrogant 'monarch of medicine' removes magic from medicine and liberates psychology from shackles of theology.

Jane and Cosmo, you live in a time when men and women are credulous, suspicious and superstitious all at once. Paracelsus (1493–1541), a swaggering Swiss physician, also lived in similar times. Yet he was the first physician to keep sorcery out of medicine.

His real name was Dr Theophrastus Bombast von Hohenheim. He was so much impressed with the works of Celsus, a first-century

Roman medical encyclopedist, that he took the name Paracelsus, meaning 'beyond Celsus'.

Paracelsus lived in a witch-crazed Europe and made little attempt to distance himself from the beliefs of contemporaries that witches were vile old women who had crooked appearance, secretive habits, anti-social behaviour and ability to fly on a pitchfork. Nevertheless, he rejected the idea that witches made pacts with evil spirits or the Devil, a fundamental aspect in the definition of witchcraft. To him witches were a physical or personality type such as the deformed, disabled, thief or murderer. He believed in evil spirits but was not convinced of their powers to interfere with terrestrial bodies. Without God's permission evil spirits were impotent. 'Dare you say that the Devil is more artful than God?' he challenged sceptics.

A witch's effect on her victims, he said, didn't come from a supernatural power, but from the feebleness of the mind of the victim. It was the psychic power of imagination, not the supernatural power of the witch, which caused discomfort, disease, even death. He suggested that the whole population didn't display the syndrome for witchcraft; this manifestation was confined to a narrow group.

I was relieved to see smile back on Cosmo's handsome face. Reading is always curative, and writing, especially about your experiences, as I have discovered, is certain to bring calmness to a depressed and disturbed mind.

'Paracelsus is absolutely right about witches' powers,' he said. 'Witches are in our minds, not around us. To weak human minds anything extraordinary or inexplicable becomes supernatural.'

'So, how would you explain the Devil's ball you saw this morning?'

He ignored me.

Paracelsus liberated psychology from the guardianship of theology when he defied the church and rejected the belief that mental illness was a sin and caused by demons though he himself believed in demons, sprites and gnomes. 'One should with diligence take note of the spirit of man, of which there are really two, that are inborn,' he said. 'For this is indeed true, that man is in the image of God and thereby has a Godly spirit in him; but, on the other hand, man is also an animal, and as such has an animal spirit. These spirits are two antagonists, and yet the one must soften the other.' If the origin of disease is psychological, he advised his fellow physicians, you should not use a medical treatment as for ordinary disease, but the treatment must be applied to the mind.

He saw the practical possibility of the power of meditation in medicine. 'If we are firm in the art of meditation, we shall be like Apostles,' he said. 'We shall not fear death, prison, martyrdom, pain, poverty, toil, hunger. We shall be able to drive out the Devil, heal the sick, revive the dead, move mountains.'

In Paracelsus' time, the practice of medicine was based on the ideas of the ancient Greek physicians Hippocrates and Galen. Some people even believed that disease came from the influence of the Devil. They sought cure in prayers, penances, exorcisms and purification by the priests, and not in potions of physicians. Paracelsus rejected these ideas and said that disease originates from external rather than internal causes, and every disease must have its remedy. 'Medicine which I had learned was faulty, and those who had written about it neither knew nor understood it,' he said. 'They all tried to teach what they did not know. So I had to look for a different approach.'

This different approach was the application of his knowledge of alchemy to search for new medicines. The true purpose of alchemy is not to make gold, he declared, but to prepare medicine. He stressed the importance of minerals in medicine, and was the first to use the processes of alchemy – the extraction of pure metals from ores – to make medicines from compounds of antimony, arsenic, mercury and zinc (he introduced zinc to the Western world). He knew some of the compounds were poisonous, but his defence would please modern practitioners: 'It is only the dosage that makes a thing either a poison or a remedy.'

'In experiments theories or arguments do not count', Paracelsus exhorted his fellow physicians. 'We pray you not to oppose the importance of the method of experiment but to follow it without prejudice.

'You are singing praises of Paracelsus as if he were a saint.' Chara frowns. 'Wasn't he arrogant, roaring drunk and quarrelsome, lived like a pig and looked like a drover and yet called himself the monarch of medicine and died from illness caused by a stab wound in a tavern brawl? What a role model for young doctors like Cosmo and Jane!'

'Forget his human weaknesses, Chara, he revolutionised medicine by establishing the importance of chemistry in medicine.'

I interrupted Cosmo again. 'You're already arrogant and you're nowhere near becoming a physician let alone a famous physician. You haven't answered my question about the Devil's ball, Dr Dorrington, sir'

'Devil's ball? You're sounding stupid, suspicious and superstitious all at the same time. If a little story can change your mind, then you have no hope of changing the outlook of people around you.'

Europe in Paracelsus' time was in throes of great intellectual revolution. A true child of the Renaissance, he laid the foundations on which the Scientific Revolution that followed his death took shape. The year 1543, two years after the death of Paracelsus, saw the publication of young Vesalius' *Fabric of the Human Body* and old Copernicus' *Revolutions*. These two books emancipated science from the bondages of sorcery. Demons were moving out of science, experiments and observations were coming in. Modern science has begun.

Cosmo scoffed. 'Thales says demons were out of science in fifteen forty-three. One hundred and fifty years on they are still clouding consciousness across the channel.'

'You're forgetting about the great Queen Elizabeth's reign which ushered in renaissance in politics, exploration, literature, theatre and even in romance.'

'What about science?'

'Great scientific lights arose in this country when Gilbert, Bacon and Harvey, all blue-blooded Englishman parched with a thirst of knowledge, came on the stage.'

'I wonder why the ignorance, superstition and false reasoning are blocking these lights and keeping everyone on this stage in an umbra. And most parts of your mind too, Miss Science. Let's do something to bring light to this town, at least. Have you got some gold trinkets for me? I'm going to see Mr Johnson tonight.'

'First tell me if it wasn't the Devil's ball, what was it?'

'Simply a ball glowing and moving around. It was mysterious, not supernatural. One day science will explain it. Let me read about the great anatomist Vesalius.'

A glorious book lays the foundations of modern anatomy and continues to influence the way we look at the human body today.

In the sixteenth century many of Galen's false ideas on human anatomy still dominated medicine. If Galen, for instance, said that the human heart had only two chambers (it has four), then two chambers was all it had.

Andreas Vesalius (1514–64), a well-off Belgian doctor's son, studied medicine at the world's leading medical school at the University of Padua in Italy. He became particularly well versed in Galen's work, but wasn't impressed by his ideas on human anatomy. Galen's knowledge of human anatomy was based on the dissection of apes, dogs and pigs, while Vesalius performed his dissections on cadavers. Like earlier professors of anatomy he didn't sit aloof on a chair while a barber-surgeon cut up the cadaver, but performed his own dissections with a touch of flamboyance, inventing instruments and improvising procedures as needed. When he became a professor at the university, most of his dissections were on corpses fresh from the gallows; the magistrates even timed the execution to coincide with his anatomy classes. From these exhaustive investigations he discovered some two hundred anatomical errors that Galen had made.

However, it wasn't easy to displace Galen. Vesalius was ridiculed when he said that men and women have an equal number of ribs. At that time people believed that men have one rib fewer than women, because of the biblical story that Eve was created out of Adam's rib. Like many scientist before him, he was obliged to defend his writings from fierce attacks. He was called a 'madman, Vesanus, whose pestilential breath poisons Europe' by his old teacher, Jacobus Sylvius, who was wedded to the teachings of Galen and refused to accept that the old master was wrong. Sylvius even suggested that straight thigh bones, which were curved according to the teachings of Galen, had changed their shape since Galen's time as a result of new fashion of wearing narrow trousers.

Vesalius was twenty-nine when in 1543 he published his findings in *Fabric of the Human Body*, a magnificently illustrated, massive book of more than seven hundred folio pages. A fortune was lavished on its nearly three hundred wood-cut pictures, which were labelled twith tiny letters so that Vesalius could refer them in the accompanying written descriptions. Vesalius wrote in the preface that the pictures 'place before the eyes of the student of nature's works, as it were, a dissected corpse'. The volume visualised for the first time in history the true structure of the human form. Its exact illustrations and dispassionate prose were based on painstaking scientific research, observation and experience. It has been called 'a glorious book, a rare and precious monument of genius, industry and liberality'.

This anatomically accurate medical textbook laid the foundations of modern anatomy. It was not a landmark in the progress of medical science, it revolutionised it.

'Salus admires Paracelsus and Vesalius but will he keep sorcery out of medicine and judge the girls' illness like them?' Cosmo sighed.

'He should read this scroll. I'll leave it on his desk.'

'Yeah.'

'When are you going to draw a witch's knot on your fireplace?'

'Surely you're joking, my beautiful superstitious scientist! I would rather untangle the knot of the noose which has now become so real.'

~~~

On most working days Salus' consulting room was crowded with patients, but a couple of days after Cosmo's encounter with the bizarre ball it was also swarmed with healthy town folks who had suddenly caught a weird disease. Dr Paracelsus would have described it as the psychic power of imagination; Dr Jane would describe it as barmybrainitis caused by a vacuous spirit known as, touch wood, superstition. Some came with a piece of paper scribbled with their accounts of how an old woman had exercised her witchcraft to cause animals to die or people to fall sick. Others sat in Cosmo's room where they mouthed off about her evil powers. Our able town constable in red coat slowly and laboriously – but sympathetically and with great conviction – inscribed their tall stories on small pieces of paper. Obviously, an enjoyable task for an aficionado of titillating tales. This high drama made Cosmo mad. 'Why did Dr Digby allow this dope to come in here? He should be at the town clerk's grubby room,' he kept on repeating under his breath. At times he was furiously grinding the pestle in an empty mortar to distract all these men whom Paracelsus would have branded feeble-minded. Other times he was boiling or mixing chemicals to create fumes apparently to disinfect spirits that had caused this feeble-mindedness.

I was dying to read these depositions. Eavesdropping had never been a moral dilemma for me though I was not so sure about sneaking into someone's room and reading their private papers. Anyway, when everyone had left I walked into Salus' room on the pretext of leaving Thales' scroll on his desk. I was greeted by a brown hedgehog that was sitting on the pile of depositions on Salus's desk. It gave me an intimidating stare with its beady black eyes and a blood-curling squeal. I was scared stiff. Witches do appear in the likeness of hedgehogs, didn't you know? I was Number One Fan of Paracelsus and showing feeble-mindedness in such a supernatural situation would have been seen as betraying my hero. Remembering his lesson on the power of meditation, I recited Lord's Prayer, took ten deep breaths and went outside to grab a stick to scare the hedgehog away. The hedgehog was nowhere to be seen when I came back with a rather very long stick; little princesses do like to keep a safe distance from ugly hedgehogs. Either my mind was showing feebleness under stress of committing an immoral act or the witch was playing tricks. I opted for the former option and bravely picked up the pile of depositions and quickly looked through some twenty pieces of papers from wooden-headed fools.

'Master Henry Byrom aged about sixty and two years of Ware testified and said that he had heard the common rumour that Elizabeth Marsh of Ware was a witch. His neighbour Master John Cannell who died last year has told him a story that added weight to the rumour. When the Marsh woman's first husband died she asked Master Cannell who was then a grave digger for her husband's skull. She informed him that he wanted to make a drinking cup out of it. Master Cannel refused to give her the skull even when she offered him ten shillings. Such an honest Christian he was. Master Byrom testified this before me the parish constable Steven Bowes as his own testimony, this 14th day of January 1684'

'Goodwife Mary Geale aged about forty and Goodwife Phillipa Milward aged about forty and six both of Hertford testified and said that about June last they were walking in the wood when they saw Elizabeth Marsh of Ware and some other women with lighted candles in their hands. It was broad daylight. Goodwife Milward counted the women and said with great fear and trepidation that they were thirteen. When Goodwife Geale showed her ignorance Goodwife Milward admonished her for not paying attention to Vicar's sermons. She said that all good Christians know that at the Last Supper the thirteenth disciple betrayed Jesus. Witches worship Antichrist and always appear in groups of thirteen. Both women said that the day they saw the said Marsh practising witchcraft was Friday the thirteenth ...'

I quickly put the papers on the desk when I saw Cosmo coming in. He remarked sarcastically. 'Does your mother know about the sins of her goody-goody daughter? She knows a lot about imagined sins of an old woman.' I felt like a she-devil, in sackcloth and ashes, with her tail under her legs.

~~~~

Bad news has wings and there are no good days to clip them. Absolutely true. The next morning the town was buzzing with the news that last night Sir Henry had signed the warrant for Mrs Marsh, and Mr Bowes was on his way to Ware to arrest and escort her to the gaol in the old castle. Mama was beaming. 'Good man he is, knows how to uphold the justice. Has great respect for your father. Immediately signed the warrant when Mr Bowes told him that Dr Digby had seen the depositions ...'

I didn't have the stomach for this depressing monologue. I walked out of the room, leaving behind the requisite minimum audience for monologues. Without Martha's presence Dr Digby would be worried about his wife's mental status. I was worried about the state of mind of my Cosmo. How would he cope with the shock of this terrible news? I wished there was a scroll from Thales to take the load off his mind.

~~~

Doctor Copernicus sets the earth in motion, but the Church stops it because the scriptures have commanded it to stand still.

'The sun is at the centre of the solar system, fixed and immobile, and planets move around it in perfect circles in the following order: Mercury, Venus, Earth with its moon, Mars, Jupiter and Saturn,' wrote Nicolaus Copernicus (1473–1543), a Polish astronomer, physician and theologian, in his book *On the Revolutions of the*

Heavenly Spheres. 'And so the sun, as if resting on a kingly throne, governs the family of stars which wheel around.'

In this book, Copernicus rejected the ancient and commonly held wisdom that the earth stood still at the centre of the universe and that the sun spun around it. To the contrary, he declared that the moon spun around the earth and the moon and the earth together spun around the sun.

Not only did Copernicus invent the idea of the solar system and placed the sun at its centre, he also gave detailed accounts of the motions of the earth, the moon and the planets that were known at the time. He said that the earth rotates on its own axis once a day, which accounts for days and nights, and revolves around the sun once yearly, which accounts for seasons.

'Thales, there is something paradoxical about Copernicus' idea that the earth moves around the sun,' says Chara. 'Not only we speak of the sun rising and setting but we actually see it rising and setting.'

'He proved our common sense wrong.'

Copernicus completed his book in about 1530, but decided not to publish it. He knew that the book would be seized and destroyed by the Church as it had emphatically displaced man from the centre of God's creation. However, his ideas had become known. Uneducated and pig-headed people merely poked fun at them; learned people, especially religious leaders, were full of venom. Martin Luther denounced him as 'an upstart astrologer, the fool who wanted to turn the whole science of astronomy upside down'.

John Calvin quoted, as a higher astronomical authority, Psalm 93 against him, 'God has established the world; it shall never be moved', and asked, 'Who will venture to place the authority of Copernicus above that of Holy Spirit?'

Copernicus passed his life in agony. His efforts to convince the Church of the truth were in vain. At the very end of his life, his pupil and friend Georg Rheticus took the manuscript to Nuremberg and printed about six hundred copies. Andreas Osiander, a Protestant theologian and one of Copernicus' students, who was supervising the printing process, was so frightened of the revolutionary ideas contained in the book that he surreptitiously removed the preface written by Copernicus and substituted his own preface in which he claimed that the book's 'hypotheses are not necessarily true; they need not be even be probable'. This colossal fraud was discovered decades after Copernicus' death.

The first copy of the printed book was sent to Copernicus in Poland. It arrived but a few hours before his death by cerebral paralysis on 24 May 1543. The copy was placed in his hands. The greatest astronomer of his time died without knowing that his book, a monument of scientific genius, had been published.

The book was in defiance of the Bible, which said 'the sun stood still', but it was too late for the Church to do anything. However, the Church duly declared it 'a false Pythagorean doctrine contrary to the Holy Scriptures,' banned it and placed it on its Index of Prohibited Books. But it was too late; the book had dealt a final blow to the earth-centred universe of Ptolemy. Astronomy would never look back.

A few decades later the Dominican monk Giordano Bruno was burned at the stake and the great Galileo was caged in his house. Their crime? Pushing Copernicus' heretical views zealously.

'Unlike them, Copernicus had the prudent good sense to die early,' Chara quips.

'Did you know Copernicus also proved that you are not unique, Jane?' Salus asked. 'You're the same decaying organic matter like everyone else.' Cosmo and I had not noticed him; he was standing near the door with a glass of cognac in his hand. We were surprised to see him in his jovial Sunday afternoon mood on Thursday evening.

'What do you mean, Salus?' I asked.

'He showed that reality is not as appears to us. When he demoted the earth to an ordinary unprivileged place in the cosmos, he showed that one's location is not likely to be special. To put it bluntly, wherever or whenever we are, it's nothing special.'

'So, the idea we humans aren't at pinnacle of God's special brought Copernicus the disfavour of the Church?' Cosmo asked.

'Absolutely.'

'What idea has brought Mrs Marsh disfavour of the Church?' I asked.

'What made you think that the Church is involved in her arrest? Witchcraft is a crime according to the statute of sixteen-o-four. The law was drafted by eminent jurists, not priests.'

'Why is then Vicar taking so much interest in Mrs Marsh if it were not a religious matter?'

'Because he believes witchcraft harms his parishioners, morally and physically. The assizes court will sit in Herford in March. Her fate will then be decided according to our laws, not the whims of individuals. '

'Would you be asked to appear as a witness, Dr Digby?'

'I do not know. I assure you Cosmo, if I'm asked to appear before the court my testimony will be guided by my knowledge of medicine, not my friendship with Vicar.'

Cosmo was ecstatic. 'Thank you, thank you, sir.'

I wasn't so happy. 'Why did you agree for her arrest?' I snapped. 'It was a heartless decision, Salus, indeed it was.'

'How could I have ignored vox populi?'

I stamped my foot on the floor. 'I'll never accept the tyranny of the voice of the people. Why did you?'

The kerfuffle brought Mama in. She jumped at me, grabbed my hair and howled. 'How you dare to speak to your father in insolent tone, young lady? Next time I'll pull out that poisonous tongue of yours.'

I wished my Mama was a shy hedgehog, not a sly fox.

Salus responded by simply turning up the wattage of his smile to 'Big Smile with Docile Eyes', the minimum setting required for a teddy bear of a husband who couldn't ignore vox imperiosus uxor, the voice of domineering wife.

Cosmo's response was to look at Mama with a rainbow smile with enough radiance to melt any woman's heart, especially middle-aged. 'I haven't finished reading the scroll. Would you also like to listen, Mrs Digby?'

Mama replied like a lamb, 'Sure, love. Go ahead.'

An astronomer with a golden nose keeps the earth standing still, yet comes up with a golden idea that smashes the old view that stars are fixed.

The Danish astronomer Tycho Brahe (1546–1601) also had the prudent good sense, not to die early but to dismiss the Copernican system. He accepted without question the dogma that the earth stood still at the centre of the universe and the sun went around it. He brusquely dismissed the earth as a hulking, lazy body, unfit for motion, and maintained that the Copernican system violated both physics and the Holy Scriptures.

Like many scientists of his time, this red-headed man with a golden nose (he had lost most of his nose in a sword fight, which he replaced by an alloy of silver and gold) believed in alchemy and astrology. Yet he was an extraordinary observer of the sky.

His real contribution to astronomy was as an observer, rather than as a theorist. He accurately measured the position of 777 stars, a remarkable achievement considering it was done without a telescope. He also measured the movement of planets, but was unable to determine their orbits, or paths around the sun. Tycho's observations helped Kepler, who was an assistant to Tycho, to discover his planetary laws.

In the history of astronomy, 11 November 1572 is a red-letter day, for on this day twenty-six year old Tycho (he is usually known by his first name) saw a new star blazing out in the constellation of Cassiopeia, which remained immovable in its place in the sky. He watched the bright star carefully night after night; it passed through various tints of purple, yellow, red, before it gradually faded.

He called it nova and in 1573 published a small book *A New Star* in which he concluded, 'this new star is not some kind of a comet or a fiery meteor, but that is a star shining in the firmament itself – one that has never previously been seen before our time, in any age since the beginning of the world.'

He went further and suggested that stars could have a beginning, middle and end. This revolutionary idea smashed the belief of the ancient Greek astronomers that stars were fixed and unchanging.

The book was almost completely ignored by his contemporaries. We know now that he observed a supernova that was visible in the northern skies for nearly two years. A supernova is an old star, not a new star, which suddenly explodes as it blasts itself apart. The remaining matter forms a neutron star, a dead star.

'Was the Star of Bethlehem like the star seen by what's-his-name?'

I was surprised to see an intelligent question pouring out of Mama's mouth. The identity of the Star of Bethlehem had puzzled astronomers for centuries. Was it the planet Venus, a comet, a conjunction or close approaching of Jupiter and Saturn in the evening sky? I had never thought about it being a nova.

Cosmo shook his head gently. 'In ancient times almost everything in the sky was called a star of some sort. There were fixed stars, planets were wandering stars, comets were hairy stars, meteors were falling stars and novas were new stars. Mrs Digby, the Star of Bethlehem was not an ordinary star; it was a holy light that guided the three Magi to the birthplace of Jesus.'

I couldn't believe what he was saying. James became Cosmo when Salus had found that his new assistant was a keen student of astronomy. He was now behaving like a bootlicker, crawler, groveller ... and all other synonyms one can find in our great language. I saw Tycho turning up his golden nose at his answer and drowning and burning him at the same time with rustic words of wisdom: a flatterer carries water in one hand and fire in the other.

Mama, a sucker for anything that was remotely biblical, beamed her approval. 'You're absolutely right, love, it was no ordinary star, it was a holy light. The Bible says it stood over the manger where the young Child was born.'

~~~

If habit were a shirt made of iron, my habit of getting up late would be a blouse made of a barrel of cannon. As usual I arrived at the church after the service had started. The days had grown shorter, cooler and darker but the congregation at Vicar's Sunday sermons had been defying the seasonal trends and had not shown any sign of shrinking. The church was warm and the sermon fiery. I soon figured out that it was about a hapless old woman who was freezing in a dark dungeon less than a mile away.

Oozing lava from his mouth and vapours from his nostrils, Vicar waved a pamphlet. 'This is the work of Satan. It's your sacred duty to find out who has published it, who is this Judas who has betrayed Jesus. Reading it is blasphemy. Burn every copy you can find to redeem yourself.'

I looked around and saw many in the congregation with a similar pamphlet. I was sitting next to old Mr Johnson and when I stared at his pamphlet he gave it on to me and whispered. 'Haven't you seen it, my lass? It was delivered it to every house during the night. I don't know why Vicar has gone nuts. Everything written here is true.'

My heart nearly stopped when I read the pamphlet.

Time Is the Mother of Truth and Truth Is the Daughter of Time

The Holy Scriptures are the only rule of righteousness. But where is it written that witches are murderers and have power to kill or have power to afflict disease or infirmity? Where is it written that witches have imps sucking of their bodies? Where is it written that witches can fly in air and do many such wonders? Where is it written that a witch makes a pact with the Devil?

The Devil has been locked into Hell and cannot interfere in human affairs; it cannot assume the shape of an innocent person.

There are evil women, as there are evil men, but there are no witches.

No doubt, there are spirits, made of incorporeal matter too fine to be perceived by human senses. These spirits do not play mad tricks upon us; it's our mind that plays tricks.

I instruct my God-fearing parishioners not to give any depositions against Goodwife Elizabeth Marsh to the parish constable or the justice of the peace. If you have made any deposition, disown it without delay.

The Reverend Peter Tyler, Doctor of Divinity
All Hollows Church, Hertford
9th of February, 1684

Vicar was still thundering. 'It's your sacred duty to find the person who has circulated it. This person has made pact with the Devil and should be hanged for making sacrilegious mischief of such proportions.'

'Hanged? Isn't Vicar going a bit far?' Mr Johnson whispered. He put his finger on his lips. 'My lips are sealed. You and I know who has done it. Don't you worry they will never find out from me. Like James, I hate hypocrisy.'

I couldn't believe what he was saying. Living a life of luxury built on bribery he seemed more honest than the man in the pulpit. I held his wrinkled hand and said in a trembling voice, 'Very much appreciated, sir.' It was mighty heart warming to find out that there was someone in the town who supported Cosmo's cause. Thank goodness!

Why had Cosmo played this senseless prank? Why the fool had taken on Vicar? Had the little devil made pact with the Devil? Who wrote the pamphlet? Thales? No, no, he wasn't mad like him. For sure he wasn't in the church, where was he then? My tiny brain was exploding with questions and more questions.

~ ~ ~

Later in the day I found him at his favourite place of solitude – a large barn with cathedral-like dignity – lying on bales of corn and looking at a scroll with the magnifying glass he had borrowed from Salus. 'You were not home when I picked it up from your front door. Thales must have left it after you had gone to the church. I know now who has been writing these scrolls.'

I sat close to him. 'Really?'

'Mr Bowes, your favourite constable. He fits the bill perfectly. Loves reading. His spidery scrawl is as laboured as Thales' mirror writing. Goes for a walk very early in the morning and has the opportunity to leave the scroll at your door without being seen by anyone. There is only one catch. I suppose if he were Thales then his stories would be about scientists frolicking with naked women not with naked ideas.'

The frown from my face disappeared and I laughed so loudly that my eyes and cheeks reddened and tears came from my eyes. My tears of joy turned into tears of sorrow when I heard Cosmo reading.

A scholar is hunted from land to land by religious hounds, imprisoned for eight years and then burned alive at the stake. His crime? ...

9 February 1600. It is freezing cold in the vast and ornate Hall of Inquisition at Rome. Fifteen illustrious cardinals of the Holy Office are seated on high-backed plush chairs forming an arc around the accused – a 51-year-old, small, thin man with black hair and dark brown eyes, kneeling silently. The Grand Inquisitor, Cardinal Severina, reads the charges. The eight counts of heresy include belief in the movement of the earth and in an infinite universe filled with innumerable worlds. Severina asks the man to recant his belief and pray for mercy to God. The man remains silent. Severina excommunicates the heretic and sentences him to die 'without the shedding of blood' (in other words, to be burned alive at the stake). The man lifts his head defiantly and declares: 'Perhaps you, my judges, may be more afraid to bring that sentence against me than I am to accept it. The time will come when all will see as I see.'

He is given eight days' grace to recant and deny his beliefs. His belief in the truth remains unshakable.

17 February. The man lies naked on a rack, his ankles and wrists bound tightly, in a dreary and damp dungeon in the terrible Tor di Nona (Tower of Nona) in Rome, which has been his home for the past two years. In the feeble light of winter dawn he is asked by the guards to put on the sulphur-coloured garb of heresy, covered with pictures of devils and crimson flames and crosses. He is then led in

chains through a howling, fanatical crowd to the site of the execution, a public square called the Campo de' Fiori (Field of Flowers). He walks calmly and with dignity, his face serene, his head high. He is stripped naked and a shirt of pitch that extends from his waist to his feet is put over him so that he would not die as quickly. The executioner ties him to the stake, piles firewood, charcoal, kindling up to the chin and places a torch between the feet. As the flames blazes around him, a priest pushes forward and presses a crucifix into his hands, but the man turns his head away. Within seconds flames sear him, smoke and fire surround him. When the fire subsides, his remains are powdered and blown in the wind so that no relic of the heretic would survive.

Someone has said that to know how to die in one century is to live for all centuries to come. The man, Giordano Bruno (1548–1600), lives on.

'Phew, what a life!' I said.

'Nothing has changed in eighty-four years,' Cosmo said in a bitter voice. 'Same dungeons, same executions in front of a cheering crowd. You and I not any different from indifferent people around us. You know Mrs Marsh has been in the gaol for three weeks and we have not even bothered to see her. '

'You're insane, Cosmo, you're insane. Printing that pamphlet was a moronic idea. Now you want to go to the gaol and meet Mrs Marsh. You have no excuse to be at the gaol. Vicar has his spies everywhere. His faithful parishioners would tell him immediately you set foot in the old castle. Vicar is not as dense as you are; he would immediately suspect that you are behind the pamphlet.'

'Who cares about him?'

'I do. Otherwise I'll never be able to say *I do* to someone I love. I am longing for that dream day in the church. Forget about Mrs Marsh. She we'll be alright. She will be judged according to our laws. Our laws are moral; we are English, not barbarians.'

'It's morality based on the false interpretation of the scriptures. It's morality that has no regard for reason. One becomes immoral when one makes judgments based on personal happiness. You know what do I mean, Miss Jane?'

My Cosmo had never addressed me as Miss Jane. I was devastated by his cool voice and detached demeanour. I cupped my face in my hands and cried.

He stood up, waved the scroll in my face and continued his monologue. 'Haven't you learned anything from these scrolls? Truth is important to scientists and they willingly die in search of the truth. The truth is that there are no witches. I'm willing to die to save Mrs Marsh from the gallows.' He lifted my face and kissed me clumsily on lips. 'I cannot do it alone, we are in together, my beautiful Jane.'

When it comes to emotions, I'm a sponge. Within seconds Cosmo's kiss squeezed out old sad thoughts, and soaked me again with happy new ones. I placed my head on his shoulder. 'I'm all yours, including my examining mind and faltering heart.'

He laughed. 'So what's your latest theory on eavesdropping, Miss Thales?'

'Eavesdropping bears fruit in all seasons. I have overheard Mrs Gardiner telling Salus that the girls are still having fits and seizures. Vicar was expecting their condition to improve after Mrs Marsh was placed in the gaol. I do not know why?'

'It's an old belief: if the witch is imprisoned, she is void of hurt, and Satan leaves her. Therefore, imprisoning the witch cures possession. If the girls are still possessed, they weren't bewitched by Mrs Marsh.'

'Will they release her now?'

'The law is not based on popular belief. She will still be tried.'

'Oh, no.'

'Gullible people also say that killing the witch cures possession.'

'Forget about what gullible people say. Listen to what this gullible girl says. I will go to the gaol with you.'

He kissed me again. 'We'll go incognito.'

'What do you mean?'

'We'll dress up in such a way that no one will recognise us.'

'Great. I'll go only if you dress up as a gypsy.' My mind was as light as a cloud. I burst into song:

The gypsy rover came over the hill
Down through the valley so shady,
He whistled and sang till the greenwoods rang
And he won the heart of a lady.

Cosmo ignored me and continued reading.

... He said the scriptures were never intended to teach science; they were not authority on the nature of the physical world. Examine before you believe.

Born seven years after the death of Copernicus, in Naples, Italy, Bruno entered the Dominican Order when he was fifteen years old. At twenty-four he fled his monastery because of suspicion of heresy and began travelling around Europe. During his travels he learned about Copernicus' book which had been published five years before his birth. He became an instant convert to the Copernican doctrine.

The runaway heretic was hunted from land to land by fanatic-hounds of the Inquisition and arrested in Venice. He was imprisoned for six years in the infamous Piombi at Venice and for two years in the Inquisition dungeons at Rome, without books, pen-and-paper or visit by friends, before he was burned alive.

Bruno's most important work on the Copernican system is *The Ash Wednesday Supper*, published in 1584, in which he describes Copernicus as 'a grave spirit, meditative, penetrating and mature; a man who did not surrender himself to any past astronomers.' Bruno also did not surrender himself completely to Copernicus. In *The Infinite Universe and the Worlds*, published in the same year, he diverges from Copernicus. This book, which started his fatal dispute with the Church, is in the form of a dialogue between two philosophers named Burchio and Fracastorio. When Burchio asks, 'Then the other worlds are inhabited like our own?', Fracastorio replies: 'If not exactly as our own, and if not more nobly, at least no less inhabited and no less nobly. For it is impossible that a rational being fairly vigilant, can imagine that these innumerable worlds, manifest as like to our own or yet more magnificent, should be destitute of similar or even superior inhabitants.'

The last decades of the sixteenth century were a period of crisis for religion and for science. This crisis dragged religion into superstition and science into scepticism. Bruno insisted that the scriptures were never intended to teach science, but morals only; and they cannot be authority on the nature of universe. He looked at the problem of the relations between God, the universe and man from a different point of view. In this view man was no longer at the centre of the universe; man's place in the universe became a minor incident in the history of an insignificant planet. This view clashed with the powerful and dogmatic Catholic Church whose motto was 'do not examine, only believe'. The Church feared that Bruno's view would threaten the idea of heaven lying just beyond the sphere of fixed star and would be dangerous to people's faith.

Copernicus had drawn a new picture of the universe, but his universe was still bounded. Bruno brushed away those boundaries and boldly predicted the existence of innumerable inhabited worlds in a universe which has neither centre nor limits. The idea of worlds beyond our own planet raised theological questions such as: Did the human beings in other worlds need redemption? Were there to be innumerable appearances of Christ? These questions embarrassed the Church.

'This life, universal, omnipresent, infinite, is the Universal Being whom men have called God,' he said. He became the apostle of new science: matter and spirit are conjoined and by studying material objects, our physical and natural environment, it was possible to read the language of Nature, and then to learn the thoughts of God.

'When we respect our planet we come closer to God,' Cosmo mumbled. 'A barren earth cannot be the spiritual home of man, sorry Jane, men and women.'

Imps

'Are you the gaoler here?' Cosmo asked in a husky voice modulated to imitate the accent of a London toff.

'Yes, my Lord,' the scruffy little man bowed and replied, visibly nervous by Cosmo's confident voice which, like all posh voices, reeked of dosh, deceit and drama.

Cosmo and I were at the old castle dressed up like a lord and a lady. The old castle was, well, old and abandoned. Now home only to bats, rats and tramps, except a row of decrepit rooms that served as the local gaol. In our chic clothes we were like peacocks in a filthy paddock. Cosmo was wearing a full-skirted velvet coat with deep buttoned-back cuffs over an embroidered waistcoat, tight breeches and leather boots. A long silk cravat around his neck, a periwig on his head and a false moustache had turned him convincingly into dashing young Lord Parker. Lady Parker was wearing a silk-and-brocade loose mantua gown over a decorated stay, the front of which was visible. The matching high-heeled silk shoes fastened with buckle and strap made me look as tall as Cosmo. I also wore a black satin half-mask to give myself an air of mystery and to conceal my identity from the town's churchgoing citizens, which included almost everyone as you could be fined by churchwardens for not showing up regularly at the house of God.

'How they call you, Master Gaoler?'

'Broderick, my Lord?'

'Do you have many prisoners tonight, Master Broderick?'

'Only two, my Lord, a thief and a witch.'

'Is the woman Goodwife Elizabeth Marsh?'

Broderick's eyes widened in amazement. 'Yes, my Lord.'

'Has anyone come to visit her?'

'No, my Lord. Witches do not have friends.'

'Really! Take us to her cell,' I said.

Sensing his hesitation, Cosmo raised his voice. 'Do you know where we are from?'

'No, my Lord.'

'Whitehall.'

'The Palace!'

'Yes, you fool. Hurry up, unless you want to find yourself in the Tower.'

'The Tower of London!'

'Yes, Master Broderick, as a gaoler you should know that's the most luxurious gaol in our great king's realm,' I said trying to hide my giggle. 'You must have heard about their skill in putting burning candles in armpits. They would love to try their new German racks and French cat's paws on you.'

'You need not fear if you do what I say,' Cosmo said.

Broderick replied in a meek voice. 'Follow me, my Lord, my Lady.'

He lighted a candle and clattered slowly down the stairs to a small, dingy cell with a nine-inch barred hole for a window which was letting in cold draughts of air. They always left a hole in the wall in a witch's' cell to let her imps enter. The cell was filled with smell of dampness and ordure, and rats scurrying around. The jangling of iron chain made us look at a corner where Mrs Marsh was sitting on a pallet of straw, her right foot tied to a chain long enough to let her walk around the cell.

'Master Broderick, give me the candle and then you may leave.'

When the gaoler had left Cosmo removed his moustache and said in his normal voice. 'Mrs Marsh, this is James. I'm here to help you.'

She just stared at us with blank eyes. The lack of food and light had taken its toll. Her pale, drawn, leathery skin made her look like those Egyptian mummies. On that freezing evening I burned with anger at the plight of an unfortunate old woman in that inferno. I wondered how we could claim to civilise natives of other lands when we committed barbaric acts in our own backyard. Our prisons were indeed warrens of decay and disease. Prisoners had to depend on meagre food supplies granted by authorities and charities. Most prisoners lived on food brought in from outside by friends and families. Even those who were lucky to escape the gallows and had served their sentences could not leave until they had paid the gaoler fees for time spent in prison. Many languished in prisons for years after their innocence had been established.

Cosmo took out a flask of brandy from his pocket and gave her. 'Sip it slowly.'

She took a few quick gulps, which gave her strength to utter feebly. 'You're a good lad, James.'

'I cannot take you out of this hell, but I'll make sure you have enough food, ale and blankets.'

'Will they hang me, James?'

'I'll never let them hang you. This is my solemn oath.' He hugged her. 'We must leave now. Sip the brandy slowly, don't gulp it.'

She started murmuring. When I listened carefully, I found she was singing in a hoarse voice:

And the Devil will fetch me now in fire,
My witchcraft to atone;
And I have troubled the dead man's grave
Shall never have rest in my own.

The gaol had broken her spirit. Her weakened mind was forcing her to believe that she was indeed an evil woman, a witch.

Broderick was waiting for us outside the cell. 'I'm surprised to hear her singing,' I whispered to Cosmo. Broderick overheard me – I had found my match in eavesdropping – and said, 'The old woman always sings when I walk out of her cell, my Lady.'

'Do you talk to her?' I asked.

'Yes, my Lady. I have urged her to turn herself to God and confess.'

'Master Broderick,' I said angrily. 'Don't ever ask her to confess. You're neither a judge nor a priest. Your duty is to look after the prisoners.'

'Yes, my Lady.'

'Give her more blankets and a mattress,' Cosmo said. 'Make sure she has plenty of meat and ale. I say every day. My soldiers will be keeping an eye on you. If you talk to anyone about our visit, do you know where I will see you next?'

'Please, my Lord, don't send me to the Tower. I'll look after her like my mother.'

Cosmo took out five gold coins and gave them to him. 'That's the spirit, Master Broderick. Keep the gold. I'll be back soon.'

Broderick grinned showing his yellowed broken teeth, 'Thank you, my Lord.'

'We are travelling in disguise. Our coach and soldiers are waiting outside the town. Master Broderick, next time you will be sitting next to our coachman. We'll make you the new chief gaoler of the Newgate. There you would mint guineas in bribes. You would love that.'

The chief gaoler of London's vast and forbidding Newgate had to buy his position and the going rate then was the princely sum of five thousand guineas. My immediate concern was where Cosmo had found five guineas. The only way he could have had that kind of money was by dipping in his savings to study at Cambridge. Why was he giving away his future?

Cosmo's words had given the little man something to dream about, I worried, at the expense of Cosmo: the Newgate, a heaven for gaolers and rich prisoners and a hell for poor prisoners. Broderick bowed so deeply that his head almost touched the ground. 'Your humble servant will not disappoint you, my Lord.'

~ ~ ~

The first great scientific work in English heralds the magnetic beginnings of the Age of Science in England. No, it won't muddle your mind.

My Lord and Lady, now I'll take you to the splendid court of Queen Elizabeth where you'll meet William Gilbert (1544–1603), a distinguished physician, who in 1601 was appointed court physician to the Virgin Queen. Gilbert's discoveries in science were the starting point for the study of two scientific disciplines we know now as magnetism and electricity.

He was the first to realise that earth is itself a giant magnet. A compass needle would dip down at different angles at different points of the globe, but would point straight down at the North Pole (previously it was thought the compass pointed to a magnetic island somewhere in the Arctic region). He introduced an instrument called the inclinometer which navigators could use for determining latitudes during overcast days.

Until Gilbert proposed his theory, a magnet was a mysterious stone to scientists. Some even believed that a compass needle points to the heaven. He discovered that if a magnet is cut in half, each half becomes a magnet.

He also distinguished between magnetism and static electricity, which were considered similar phenomena. Amber produces static electricity when rubbed with a cloth. From the Greek word *elektron*, Gilbert coined the Latin word *electricitas* for this property that amber displayed. The word was soon anglicised to 'electricity'. He also invented an electrical apparatus, the forerunner of a simple electroscope, to conduct experiments on the electrical properties of various substances.

In 1600, Gilbert published a book, *De Magnete*, in which he described valuable facts and experiments on magnetism and static electricity. He proved that the Earth is a magnet and showed that an iron rod can be magnetised by forging it in the north-south direction.

Gilbert always performed his experiments with great care, and noted down every observation he made. *De Magnete*, a great exposition of the new scientific methods, is often considered to be the first great scientific work in English and the first-ever learned work on experimental physics.

He was an astute experimenter and maintained that 'stronger reasons are obtained from sure experiments and demonstrated arguments'. He was scornful of bookish knowledge of 'people who blindly trust authorities, lettered clowns, grammarians, sophists, pettifoggers and the wrong-headed rabble', and dedicated his 'foundations of magnetic science' to those 'who look for knowledge not in books but in things themselves'.

While Gilbert was enthusiastic in support of the Copernican system, the great experimenter was incorrect when he speculated that magnetism was the actual cause of the earth's rotation. But his science was on the ball when he criticised — with the full force of invectives — claims for magnetic perpetual motion machines, exclaiming: 'May the gods damn all such sham, pilfered, distorted works, which do but muddle the minds of students.'

'You know that your idea of paying five guineas to give meat and ale daily to Mrs Marsh was like Gilbert's idea of magnetism rotating the earth daily.' I said to Cosmo. 'Not a realistic idea. Never trust a lump of lodestone or a grubby gaoler to do any real work. You have wasted your hard-earned money.'

'What do you suggest?' Cosmo asked sarcastically. 'A perpetual machine or a semi-omnipotent engine to feed her?'

'What on earth is a semi-omnipotent engine?'

'A machine so contrived that it could move forward or backward, upward or downward, circulatory or corner-wise, to and fro, straight, upright or downright. That's how the Marquis of Worcester dreamed it in sixteen-sixty-three. He liked it so much that he wished that a model should be buried with him.'

'An engine carrying a basket of food to Mrs Marsh's cell each night, dodging Vicar's spiritual snoopers. That's my dream machine.'

'If life were a dream.'

~ ~ ~

I was sitting in the parlour of our house and wondering about Mrs Marsh's cat and who was looking after her when a coach stopped in front of the house and Mr Bartlett jumped out of it and furiously knocked at the door. Premonition or not, his visit turned out to be about the cat.

When I opened the door I saw Mrs Bartlett standing outside carrying Alice in her arms and crying loudly. Alice's body was covered in bruises. I was alone. Salus and Mama had gone to the church.

The dunces of Oxford and Cambridge would never allow a woman in their sanctimonious male bastion, but did they know that women would always make better doctors because they had the God-given gift of nursing sick people? I ran to Salus' room and came back in a flash with an armful of cotton dressings, potions and ointments. Mr Bartlett was speechless, dumbstruck with terror; Mrs Bartlett was still holding Alice in his arms and mumbling some psalms. I asked her to put Alice on the couch and started cleaning her wounds. I soon found out that my patient was not unconscious and the wounds were not deep. Up yours, Oxford and Cambridge, who needs your medical degree? I could teach their degree-holders a thing or two, first, throw out hysterical parents from their rooms when they are treating a child; and, second, double your fees if they refuse to leave.

As I had seen them before in a similar situation, Mr Bartlett had not spoken a word and Mrs Bartlett was hoarsely reciting the

Lord's Prayer. I had treated the wounds of my little, calm patient; now was the time to treat my big, neurotic ones. I went inside to fetch two mugs of ale. I gave some pills to them to swallow with ale.

Salus' orangeypills did the job again. Mr Bartlett had found his voice. 'Nice ale.' He emptied the mug in one quick gulp. 'We used to have the best ale in these parts but since Mr Donne died we have not yet found a good brewer. The witch has ruined our life.'

I pretended to hate Mrs Marsh. 'Don't worry. The witch is in the gaol. She can't harm anyone.'

Mrs Bartlett smiled when she saw Alice opening her eyes and quietly looking at us. She put the Bible down, which she had been clutching since she came in. Apparently, she no longer needed the consolations of the Bible. Why do the scriptures lose their lure when man is lulled into a sense of security? Man? Sorry, I spoke like those despicable dons who blithely ignored the other half of humanity and to whom men were proper 'blokes', 'chaps' or 'fellows' but women were indecent 'bags', 'bitches', 'broads', 'chicks', 'cows', 'crumpets' … a long list that ends with 'witches'.

'It wasn't to be our destiny,' she said stroking Alice's face which had only minor scratches. 'The witch is still hurting my girls.'

I consoled her. 'Her wounds will heal in a few days. They won't leave any marks on her body.'

'God bless you.'

'They look like the scratching of paws of some animal.'

'It's the witch's cat,' Mr Bartlett replied. 'We can't get rid of her. She is always hanging around Alice.'

'Do you like cats?' I asked Alice.

She looked at me steadily. Mrs Bartlett replied for her, 'Not the witch's cat. But she loves cats. Don't you, dear?'

'She is not a cat,' Alice said sharply. 'Don't you know Mother the witch can change herself into a cat?'

'I tried to kill her with a sword and she jumped at Alice and attacked her like a mad demon,' Mr Bartlett said.

Alice raised her head and looked at her father. 'Father, the cat will die only when the witch dies.'

Mr Bartlett sighed. 'She will be hanged for sure, but her day in the court is weeks away.'

Mrs Bartlett picked up the Bible, pressed it against her bosom and started crying. 'My innocent girls will suffer more. Please Lord, help us.'

Even if I ever became Dr Jane Digby, I would hate to do the work of a priest or a spiritual healer. I couldn't stand crying parents. I wanted to get rid of them. 'You should take Alice to the church and pray over there.'

My words turned Alice into the cattiest girl I had ever seen. She howled me down. 'You are a witch and a stupid bitch. Amy says that you and your boyfriend will never be doctors. She'd put a curse on both of you.'

Mrs Bartlett ignored her remarks. 'No dear, we won't go to the church.'

I gave Mr Bartlett a jar of ointment. 'Please apply on wounds twice a day. Don't dress them.'

Mr Bartlett took out some gold coins and gave to me. 'Thank you, Miss Jane. You're already a good doctor.'

Alice snarled at me. 'She is a witch, not a doctor. She dresses up like a lady and visits her friend in the gaol.'

Mr Bartlett bundled up her crazy daughter and walked out of the door, Mrs Bartlett followed him sobbing.

I counted the coins. They were five. I was overjoyed that I had recouped what Cosmo had lost the night before. My happiness didn't last long when I thought about Alice's comment. How come the little bitch knew that I had visited the gaol dressed like a lady?

~~~

Sir Henry, I imagined, would be a crusty old man dressed like a magistrate in a long black gown and top hat, but he turned out to be a cute thirty-something dandy. He took off his hat and bowed when I opened the door. 'I'm here to see Dr Digby. I presume you're his famous daughter. Vicar Tayler speaks very highly of you.'

I was surprised at his gentlemanly gesture of taking his hat off instead of simply doffing it. 'It's very kind of him.'

He put his hat on; hats were worn indoors as they provided warmth in our draughty rooms. 'Mr Bartlett never stops talking about your exceptional medical skills.'

'I'm flattered, sir. I can never become a licensed physician, you know that, sir.'

'Times are changing. Our universities will change too.'

'The dogmas of old dons never die.'

He cackled with laughter. Hearing his cackle Salus and Vicar came out to receive him. They had been waiting for him. He seemed to me a renaissance sort of man. I wished his views on witchcraft were as refreshing as his cackle and clothes.

Ignoring Cosmo's threatening demeanour I kept my ears attached to the door of Salus' room. When I couldn't hear anything I improvised by placing a pewter tankard on the keyhole to magnify their voices. You know now how to turn passive eavesdropping into an active one, both unforgiveable sins in Cosmo's cosmos. Not in mine.

Imps. Imps. Imps. That's all I kept on hearing that afternoon. The keeping of imps, or familiars, was the first proof of the witchcraft. The imps, either visible or invisible, appeared in the shape of any animal or insect. They made a satanic pact with the witch before they would do the Devil's work. For professional services rendered to the witch, they demanded to attach themselves to some part of the witch's body and suck blood out of her, leaving a mark. In Vicar's eyes, Mrs Marsh's cat was not a cat but an imp kept by her and sent out to perform evil deeds. I was disheartened to discover that neither the doctor nor the dandy discounted this loopy relationship between a woman and an animal.

Their voices came clearly through my tankard when the man of God raised the tempo of their spiritual discussion. 'Sir Henry, as a Christian and the upholder of the law it's incumbent on you to act. You most order the search of the body of that Marsh woman for any signs that she had suckled evil spirits.'

'As you wish Vicar,' the dandy replied.

'How do you propose to do it?' the doctor asked.

'Sir Henry should appoint four honest and respectable women to examine her body.'

'Whom do you suggest, Vicar?' the dandy asked.

'Mrs Gardiner and Mrs Meriwether, they are experienced midwives in my parish, if you do not know, Sir Henry. Mrs Bowes would be my next choice. You probably know our constable's wife. She has remarkable knowledge about witchcraft. The fourth woman ...'

'If I may interrupt you Vicar, Miss Digby should also be on this panel of juries. We must encourage our young people to take part in the affairs of law.'

'I absolutely agree with you. Miss Digby is an excellent choice.'

The experienced doctor protested. 'She lacks the experience.'

'She will learn on the job,' the young dandy said. 'The panel would work under your supervision. We're lucky that a doctor of your calibre has decided to practise here and not in London.'

'That's very kind of you, Sir Henry, but you should spare me from this duty.'

The man of God took out a stick and thundered. 'John, you can't abdicate your responsibility.'

The dandy threw a carrot. 'You know our College of Physicians doesn't approve physicians outside London to become Fellows. I plan to petition the king that this rule should be changed and eminent country physicians like you should be admitted as Fellows.'

The doctor bit into the juicy carrot. 'You leave me no choice.'

The dandy cackled. 'I'm honoured, Dr Digby. I'll order the gaoler to bring the woman to your rooms for examination. Send your statement to me and I shall present it to the court.'

Cosmo almost fainted when I relayed the bad news to him.

Voices were still emanating from the room so I again placed my powerful listening device on the keyhole.

Vicar was speaking to Sir Henry in an agitated voice. 'What have you done about that scurrilous pamphlet?'

'Well, nothing worthwhile has come out of parish constables' enquiries, though Bowes has stumbled on an interesting story. You know it's all based on rumours. You wouldn't be interested in rumours, would you, Vicar?'

'No, of course, not. But rumours could lead us to the truth.'

'If you do insist, Vicar. Mr Johnson told Bowes that when he was coming back from Oxford, there was a stranger in his stagecoach carrying something that looked like a bundle of pamphlets. When the stagecoach stopped at Hertford he passed on the bundle to a man dressed in a cassock and surplice. Mr Johnson has confided to Bowes that the Hertford man looked familiar, but the night was so dark that he couldn't vouch for whether the man was indeed a priest.'

The doctor expressed his genuine surprise. 'I can't imagine anyone undermining you, especially within our own parish church?'

The dandy cackled softly.

The man of God was shell shocked.

I chuckled that the corrupt and crafty old Mr Johnson had sown seeds of acrimony between Vicar and Mr Harsnet. I couldn't see through the door but I knew that Vicar was fumbling through his Bible. Didn't he know that the Holy Book doesn't have any passages that he could use to put a curse on his deacon? Only witches do such satanic things.

Upon hearing this story, Cosmo hugged me tightly. 'The best news I have heard in ages. I'm going out to see Mr Johnson right now. Tell Dr Digby that he has taken ill and I have to deliver medicines to him.'

'What medicines?'

'Red wine and chocolate, silly girl, medicines recommended by doctors for good health and hearty celebrations.'

~ ~ ~

Observations and experiments can turn a hypothesis into a law, scientifically. But no scientific law can spare you from the Tower of London.

Greetings Jane and Cosmo. I'm frustrated that my efforts to keep you, and through you, the town folks, on the path of reason have failed so far. As a scientist I have no choice but to continue my quest after truth.

I know you are also seeking the answer to a question that haunts you: Why are people around you engulfed in flames of ignorance and superstition? At the same time you take solace in the fact that there are people like Mr Johnson who have stepped out of darkness.

Gilbert took English science out of darkness. The English Renaissance continued with Francis Bacon (1561–1626), a statesman and philosopher. You must be wondering that nearly seven decades after Bacon's death, the renaissance of science has failed to change the society. Science's impact on society is excruciatingly slow. Sometimes scientists are far ahead of their time that it takes generations to understand fully their ideas and theories. Science's aim is not to change society, but to enrich it by new discoveries and invention. For their search scientists need right tools.

Bacon discovered the most important tool of science – the scientific method – but he did not make any significant scientific discovery. 'I shall content myself to awake better spirits like a bell-ringer, which is first up to call others to church', he once wrote to a friend. Bacon's bell is still ringing and, I hope, will keep on ringing forever.

Bacon advocated a new method of inquiry, completely different from the philosophical methods of ancient Greeks, in his book *Novum Organum*, which has influenced every scientist since its publication in 1620. He said that scientific laws must be based on observations and experiments.

He rejected Aristotle's deductive, or *a priori*, approach to reasoning, and suggested his inductive, or *a posteriori*, approach. 'Deduction consists of prepositions, prepositions consist of words,

words are symbols of notions,' he said. 'Therefore, if the notions themselves are confused and over-hastily abstracted from the facts, there can be no firmness in the superstructure. Our only hope therefore lies in a true induction.'

The essence of his method is as follows: collect masses of facts by observations and experiments, analyse facts by drawing up tables of negative, affirmative and variable instances of the phenomenon, draw up tentative hypotheses from the evidence, collect further evidence to verify them and then proceed towards a more general theory.

Bacon was the first to envision a scientific laboratory, a self-contained space for conducting controlled experiments. However, the term 'laboratory' was introduced by the English dramatist and poet Ben Jonson in a masque performed in 1610. Jonson's image of a laboratory was based on alchemists' workshops which typically contained an array of furnaces, bellows, cauldrons, crucibles, alembics, stills, jars, flasks and so on.

'Bacon might have envisioned a laboratory,' Chara points out, 'but he never conducted any experiments.'

It's curious that the man who gave science inductive and experimental approach and who was a contemporary of scientific greats such as Gilbert, Kepler, Galileo and Harvey didn't take any part in scientific developments with which they were associated.

On the personal side, he was somewhat vain and lived beyond his means. In 1621, when he was Lord Chancellor of England, he was convicted of accepting bribes, fined forty thousand pounds, sentenced to the Tower of London and barred from the Parliament. However, the king soon pardoned him.

Jane and Cosmo, when you read my scrolls, remember Bacon's words: 'Read not to contradict and confute; nor to believe and take for granted; nor to find talk and discourse; but to weigh and consider.'

He expected scientists to weigh and consider their ideas by experiments and observations, not authority. Aristotle challenged the authority of his predecessors, including myself. Bacon challenged Aristotle. His *Novum Organum* was a challenge to Aristotle's book, *Organum.* That's how science advances.

In my opinion, he was a bit harsh when he said the ancient Greeks were like boys 'prompt to prattle' and full of wisdom which 'abounds in words but is barren of works'. Someone has said that a science is any discipline in which the fool of the generation can go beyond the point reached by the genius of the last generation. On this criterion, we all are fools, we all are geniuses. It all depends where in history our judge is placed.

I yawned. Cosmo rolled the scroll and gently touched my head with it. 'Tell me my sleeping beauty, why was light on in your bedroom about midnight?'

'What the hell were you doing at that ungodly hour in our street?'

'Looking for witches.'

'Did you find any?'

'Yes, one hidden in a bedroom. I was hoping she would be sitting on the window sill ready to fly into my arms.'

'She was a wise witch.'

'What were you doing? Reading a romantic sonnet?'

'I'm not a fan of bards, balding or hairy.'

'What about scientists, fat and ugly like Socrates?'

I smiled. 'You would be worried if I say like cute justices like Sir Henry. Where do you think the English justices are placed in history?'

'In the year of our Lord one thousand six hundred and eighty-four and in the pigsty of superstition and ignorance; to them Bacon is simply bacon.'

'Jealous, ain't you? I have read somewhere that Bacon uses the term *trial* to describe an experiment. Tell me why?'

'He believed that in search of truth scientists are not forbidden from the inquisition of nature.'

'Like the Inquisition of Bruno? Torture nature to extract confession?'

'Torture nature or even neglect her and we humans will have no place to live. He said that scientists should design experiments in which nature is put to the question. He was a lawyer and uses lawyers' technique to examine the works of ancient Greeks. In one of his books he summons Aristotle, Plato, Galen and other *sham philosophers* to the bar to face persecution for *vague inductions* and *bogus cures.*'

'Could you explain this method to my father so that he also arrives at the truth?'

'What do you think is the truth?'

'That Mrs Marsh doesn't have any Devil's marks on her body.'

'Well, first he has to observe, but he can't observe directly. Unfortunately, he would have to rely on second-hand observations of four women. From their information he should form a hypothesis. He should then inquisition the women Bacon-style to test his hypothesis. The next stage is the formulation of theory. His statement should be based on that theory.'

'A right prescription for Dr Digby.'

'I can't prescribe your father to follow the scientific method, but I can ram it into his daughter's head. Repeat after me: the scientific method is a continuous interplay of observation and hypothesis – observations lead to new hypotheses, which guide more experiments, which help to change existing theories.'

I yawned again. 'That's all?'

Cosmo stretched and yawned. 'Stop yawning. Yawning is contagious. More than half of the people yawn within five minutes of seeing someone else yawn.'

'Your theory is one big yawn.'

'A theory becomes a scientific law when it has been verified mathematically or empirically. A scientific law could be used to predict the behaviour of nature, including yawning.'

'I suppose there is no scientific law that could predict the behaviour of the crown court?'

'But its behaviour could be influenced. I have got an idea, Jane?'

'No, no, Cosmo, not one of your dangerous ideas.'

'We should print a pamphlet about the scientific method and send copies to the judges and all members of the jury.'

'No, no, you are not going to do it. Next you would start bribing members of the jury five guineas each. Good ol' Mr Johnson can't save you again. Count your blessings and forget about it.'

'Is that the order from Queen Jane, She Who Must Be Obeyed?'

'Yes. Bribing or influencing the jury is a damn stupid idea. Like Bacon you could find yourself in the Tower of London. The king won't remit your sentence. I'll then be Widow Jane, She Who Never Married.'

~ ~ ~

Mrs Gardiner showed us Salus' magnifying glass. 'You may use this to examine her body. Dr Digby has warned that when you are looking through this glass and if a mark looks as big as a quarter of a penny, it's not really as big as that. In your deposition you should write the size of the mark as you see with the naked eye.'

We, the proud jury of three puddingy matrons and one skimmed-milk lass, were assembled in a back room in our sprawling house waiting for the gaoler to bring in Mrs Marsh. Salus had been methodical and, I hoped Cosmo would permit me to add the word, scientific. He had asked us to examine the prisoner as a group and individually. He had given each of us a piece of paper to write our statements. 'Soon after you have completed your examination, each of you would write your statement, sign it and give it to me,' he had warned. 'You can discuss your observations with each other, but you can't show your depositions to others. If anyone tries to influence you to change your views, please tell me confidentially. I shall request Sir Henry to charge that person with perverting the course of justice.' I felt proud of my father; such surges of pride were rare, I must say.

Mrs Bowes took the magnifying glass from Mrs Gardiner, looked through it and when failed to focus on any object, grumbled, 'This newfangled gadget can't help you if you do not know what you're looking for. Did you know how the Devil leaves marks on a witch?'

No one replied. She grabbed the opportunity to undermine Mrs Gardiner as the leader of the group. 'The Devil leaves small or big tits upon the body of a witch when he sucks the blood. He leaves other marks when he kisses, licks or touches the body of a witch or rakes his claws across the body.'

I raised my hand to ask a question as if I was in my old classroom. 'What are imps, ma'am?'

'You shouldn't be here if you do not know what imps are. They are the Devil's agents.'

'We know one of her imps is a cat,' Mrs Gardiner said.

'Is it true, Mrs Bowes, if a witch has one mark, she has one imp? Mrs Meriwether asked.'

'You're absolutely right.'

I raised my hand again. 'Where do these bruises and marks appear most?'

'You will find them under her armpits, under the eyelids, within the lips, on her breasts or in her private parts. Don't be shy to examine her private parts, young lady.'

Mrs Gardiner deflected her sarcastic comment. 'Don't you worry about that, Mrs Bowes. Miss Jane has much experience of examining women patients, more than you do.'

Mrs Bowes was quick to retort. 'Not of examining witches.'

My hand was up again. She hadn't yet realised that I was taking the mickey out of her schoolmarmish style. 'What about warts, calluses, moles and other natural marks?'

She took out a needle from the sleeve of her blouse and displayed it like an important medical instrument. 'Prick the mark with this sharp needle. If it's the Devils' mark, it won't bleed and the witch won't feel any pain.'

I felt like shouting 'it's all bullshit' but restrained myself. Swearing and cursing was unlawful and the penalty was one shilling. All Sir Henry needed for convicting me was two witnesses; I was surrounded by three sourpusses.

Our lesson on the witches' teats and marks ended when Salus knocked on the door accompanied by Mrs Marsh. I was delighted that he had asked Broderick to remove her chains. She looked better than I had seen her in prison. The grimy gaoler was undoubtedly in awe of Lord Parker, still dreaming of a ride in his coach to the Newgate.

When Salus had left Mrs Gardiner stripped Mrs Marsh, shaved her head and private parts and asked her to stand on a stool. Even from a distance I could see that there were no bruises, marks or teats on her body, though a small mole near her private parts made me uneasy. Even without the help of the magnifying glass, to Dr Jane Digby it was indeed a mole, but what a fanatical matron and two midwives who were close to the bewitched girls' families would make of it. I worried that they could make a mountain out of a molehill. I wished I could show that mole to Salus.

The moment Mrs Bowes spotted the mole, she mutated into a wolf and growled. 'In what form the Devil has come to you?'

Mrs Marsh mumbled incoherently.

'What shape does he take when he comes to you?'

I was shocked to see her mumbling. 'As a little dark-haired man.'

'Did the man have cloven feet?'

'Hard feet.'

'When did he come to you first?' Mrs Meriwether asked.

'Last night.'

'What did he say to you?'

'Bess, I must lie with you.'

'And then?'

'He pissed on the floor; Bess, I have sprinkled you with holy water, you are now a witch.'

She was confused and drunk; perhaps the little dark-haired gaoler had given her too much ale. I went closer to her face to smell her breath. It did smell of ale.

Mrs Bowes took out a needle and pricked deeply into the mole. 'Did the Devil suck you there?'

She was probably too numb to feel any pain and mumbled something. Accustomed to her mumbling I could hear her singing.

She said something even stranger,
Which causes me great horror:
That the Devil became a man
And took ardent lustful pleasure
With her. Oh, God, what horror!

Mrs Bowes demonstrated like the great anatomist Vesalius used to demonstrate dissections to his university students. 'Everyone look at this mole. She feels no pain and it's not bleeding. Do you know what does it mean?'

Freshwomen Gardiner and Meriwether replied in chorus. 'It's the Devil's mark.'

Neurons in their superstition-saturated brains started sending signals to superimpose the stereotypical image of an ugly old hag wearing a black cape and black pointy hat on a sweet old dame wearing nothing but dusty skin, her glabrous head glistening, her dirty red hair lying near her feet. So much so for Salus' scientific approach. Science's curtain again failed to cover up superstition.

~~~

Doctor Harvey explains the circulation of blood, ignores the critics

...

The blood, pushed by the heart, moves as it were, in a circle. The heart is a pump made of muscles. It causes the blood to circulate throughout the body through the arteries and then back to the heart through veins.

We now take this for granted, but until the time of William Harvey (1578–1657) people were not aware of blood circulation. It's surprising that Vesalius' profound knowledge of the human body didn't extend to the heart.

People believed that the heart was a kind of furnace for manufacturing 'spirits' necessary for many part of the body. Since blood is bright red in the arteries and bluish in the veins, they believed that there were two distinct systems of blood flow in the body. They also believed that blood was made in the liver and flowed through the septum (the dividing wall) of the heart before being absorbed by the body.

For years Harvey dissected many animals to observe the functioning of the heart. There was no microscope in those days and his magnifying glass was not powerful enough to see capillaries. Yet, from these experiments he concluded that the heart is a pump made of muscle. It has four chambers: two upper auricles and two lower ventricles. The blood is pumped by the left ventricle through the arteries, and it returns through the veins to the right auricle. It then passes through the right ventricle which pumps it to the lungs, where it changes from a bluish to a bright red colour. It returns from the lung to the left auricle and then passes through the left ventricle. Thus it circulates continuously.

Harvey waited for years before publishing his findings in 1628 in his book *De Motu Cordis* (Movement of the Heart and Blood in Animals); he was afraid that they wouldn't be accepted as 'respect for antiquity influences all men: still the die is cast and my trust is in my love of truth'.

Harvey was Bacon's friend and physician and acquaintance of John Aubrey, the gossipy biographer of the time. Aubrey, who described Harvey as 'a little man and very chloric', has left an account of him in his *Brief Lives*.

When Harvey published this theory, it was rejected by his fellow physicians because they maintained that his ideas had not cured a single patient. Aubrey writes, 'I heard Harvey say that after his book came out, he fell mightily in his practice. 'Twas believed by the vulgar that he was crack-brained, and all the physicians were against him. I knew several doctors in London that would not have given threepence for one of his discoveries.'

He was ridiculed as a *'circulator'* (the Latin slang for 'quack'). 'All the physicians were against his opinion, and envied him; many wrote against him', notes Aubrey. 'But in about twenty or thirty years of time, it was received in all the universities of the world.'

It took nearly half a century before Harvey's discoveries, which were the result of painstaking experiments on animals of kinds, were accepted by the medical profession. The new generation of medical students were no longer tied to the teachings of Hippocrates and Galen, they were now exposed to the pioneering works of Paracelsus, Vesalius and Harvey.

... and examines superstition in a scientific way.

Jane and Cosmo, you would be surprised to know that exactly fifty years ago in 1634 on the order of King Charles a jury of midwives was appointed to examine the bodies of four women accused of witchcraft. The order said that the 'said midwives are to receive instruction from Mr Dr Harvey his Majesty's Physician'.

This incredulous tale of witchcraft began when eleven-year-old Edmund Robinson, the son of a woodcutter living on the borders of Pendle Forest in Lancashire, made an elaborate deposition before two justices of the peace. He said that as he was roaming in the forest picking blackberries he saw two greyhounds, one brown the other black, running in his direction. There was no one with them and he thought they belonged to two of the gentlemen living in the neighbourhood. The dogs came to him and fawned on him. He noticed that each dog had a collar which shone like gold. A hare appeared at the same time. He cried, 'Loo, loo, loo', but the dogs wouldn't chase the hare. Angry at the laziness of the dogs he took a stick and was about to beat them when the black dog stood up as a woman and the brown dog as a little boy. The woman was a Mrs Dickinson who lived near his house. She offered him a silver coin to keep quiet about it. He refused the bribe saying, 'No, you're a witch.' She then took a bridle out of her pocket and jingled it over the head of the little boy turning him into a white horse. Mrs Dickinson seized Johnson in his arms and mounted on the horse. They rode with inconceivable swiftness over forests, fields, bogs and streams until they came to a large barn called Hoarstones, a locality well known for witches' sabbath, a meeting of witches, where he saw many old women. Johnson's father endorsed his deposition saying that he had gone to look for him and found him in a state of terror.

In his deposition, Johnson said he could identify the women present at the Sabbath. His father led to him from church to church to identify these women. The boy received money for each one he recognised.

As a result of this witch-hunt nearly thirty women were arrested. After they had been searched for Devil's marks, seventeen were tried and seven found guilty. Three died in the gaol. When the affair came to the attention of King Charles, who unlike his father King James was not convinced of the reality of magic and witchcraft, ordered Harvey to supervise the examination of the four women. The women were brought to London and their bodies searched for any teats and marks by a jury of nine midwives. In his statement Harvey concluded that 'nothing unnatural was found, neither in their secrets nor any other parts of their bodies'. The four women were pardoned by the king.

Subsequently, Robinson was brought to London with his father and was re-examined alone. He confessed that he had made up the whole story up. His father had taught him what to say in the hope of making money.

There is another story of Harvey's interest in looking at superstition in a scientific way. Once Harvey disguised himself as a wizard and went to a lone house where an old woman reputed to be a witch lived. He asked her to show her imp, a toad. She poured some milk in a dish and whistled. A toad came out from under a chest and drank some of the milk. Harvey gave a shilling to the woman and sent her away to buy some ale. He then opened up the belly of the toad with his dissecting knife and out came some milk. He concluded that the toad was not a devil but a mere animal.

These stories show the advance of science over superstition, but this advance so far has been patchy.

'Grass is always greener on the other side of the fence,' Cosmo groaned, munching a chocolate. 'This side has only weeds.'

'What makes you think Salus won't follow the example of Harvey? Wait till he reads it.' I picked up the scroll and opened the door to his room to leave it on his desk. Mama was leaning on the desk reading a paper. 'Sneaking in to read your father's deposition, are you?'

'I'm not a curious cat, Mama.' I replied, leaving 'like you' hanging on my lips. 'I came in to put this on his desk.'

'Well, you'll never get a chance to read it.' She smirked. 'I'm about to deliver it to Mr Bowes. He would take it to Sir Henry tonight.'

Her words were like a bitter pill that I had to swallow. Why wasn't life like orangeypills? Phoney they might have been, yet full of hope and promises. I craved for a chocolaty kiss from Cosmo to pull me out of the deep ditch of disappointment.

Coven

Where was the first cuckoo of spring to sing the end of the winter of endless lies? Where was the bright sunlight to brush away snow and paint the woods green? In driving rains and snows I was yearning for Cosmo's balmy warmth. He had been away for almost a week; his mother gravely ill.

Mama poked her head through the door of my room. 'Has your father come back?'

'He is with patients in his rooms,' I replied absentmindedly.

'Didn't you know he has gone to the courtroom?'

'Judges are not here until tomorrow.'

'He has been called before the grand jury.'

'Blimey!'

While worry clouded my face, Mama was on cloud nine; she had been herding the women putting on Vicar's feast for the assizes judges. Every year when primroses would begin to bud in our garden, two judges would be dispatched from the courts in Westminster to dispense justice in criminal cases in our town. Their Lordships were not immune from the acts of crime, for the roads abounded with bandits and highwaymen; and if they escaped their violence, petty thieves lusted for the pleasure of stealing from judges and lawyers who often accompanied them. But the town folks treated the judges with great pomp and ceremony. As they reached the border of the town, they would be welcomed by the gentry, the mayor, bailiffs, sheriffs, liverymen and constables, and then parade through the town to the church to soak up righteousness and stuff down delicacies.

'I'm off to the church again,' she said and disappeared. She had obviously popped in to find out the result of the committal hearing. The saucy news of a witch trial would be a spicy garnish on the dishes the women were cooking. Witches' trials were like fairs, people looked forward to attending them. Spectators crowded the courts. Rumour mills worked overtime.

Why had the grand jury called Salus? I needed to talk to someone, anyone. The only person I could think of willing to lend a sympathetic ear was Mr Johnson. I rugged up in my fur coat and shouted a lie, 'Martha, I'm going to High Street to buy shoes.' I didn't wait for her reply; she would be asking hundreds of questions. Shoes not only touch the soles of women's feet but also their souls.

Mr Johnson hardly knew me but he treated me warmly like his long lost granddaughter. He made me sit comfortably in a big leather-covered chair in his wainscoted sitting room, placed a cashmere rug over my feet, stoked the marble fireplace, opened a walnut bureau-cabinet, picked up a silver goblet, filled it up to the brim with cognac and gave it to me, his hands shaking. 'Drink it up, lass, you need it.'

I realised that the old fox had a heart of a young pup. I sipped the cognac. 'Thank you, sir.' I had never seen him smiling. When

I looked at him closely I knew what had caused his fixed facial expression. It was some affliction I didn't know about, not the meanness of his heart.

He drawled. 'Bad weather. Sad times. I guess you have heard the news?'

'What news?'

'The grand jury has committed the old woman to trial.'

I sighed. 'Oh, my God.'

'Didn't your father tell you?'

'I haven't seen him whole day.'

'They have indicted her on many counts, including bewitching Mr Donne to death.'

'They were quick.'

'Rumour has it that the jury sat only for less than hour. Read the depositions and rubber stamped Sir Henry's recommendation. Bloody bastards! Even a coven of witches wouldn't dare to do such a dastardly deed.'

'Pff!'

'The bastards had the gall to say that they hadn't indicted her for being a witch, but for invocation of some spirit to harm people.'

The members of the grand jury were invariably well-off gentlemen (no women, gentle or otherwise), and their decisions reached by simple majority.

'Did you know who the members were?' I asked.

'Seventeen bloody toffs as fake as a three-shilling coin. Who cares who they were?'

'Did you hear anything about my father's deposition?'

'No. I guess Dr Digby must have supported the allegation. I suppose they called him to verify his statement. The bastards wouldn't have the intelligence to challenge him.'

'Pff!'

'You know the poor woman would be hanged if found guilty.'

'When is the trial?'

'They say it's on Wednesday.'

I heard a tiny voice, 'Francis, you in there?', then walked in the owner of the voice. Mrs Johnson was surprised to see me.

'Nice to see you here, love,' she said and hesitatingly looked at her husband.

'What's the matter, gentlewoman?'

'Bad news, I'm afraid.'

'No, no, it's not about James' mother?' I asked.

'I'm sorry, love, she passed away peacefully this afternoon.'

Tears welled in my eyes. The old man's face twitched. The face that couldn't register smile instantly turned sad. The old woman hugged me. 'James will be alright, love. We'll look after him like our son. He's a good lad.'

~~~

The funeral turned out to be a big affair, almost everyone was there, including many from the landed gentry. Mr Johnson went from house to house, walking with the help of a stick, his body trembling, leaning every now and then on his wife's shoulder, coaxing everyone into attending the funeral, saying 'the orphan needs our support in his hour of need'.

After the funeral Vicar invited Salus and Mama to his house to hobnob with Sir Henry and his wife. Their heads light and their tummies heavy after partaking sweetened wine and yummy pies, all paid for by bribes collected by an old man decades ago, they were perhaps hoping to collect dividends on the bribes they had given each other: the tormented dandy spiritual blessings, the ambitious doctor a dream fellowship and the wicked man of cloth the naked power.

When I returned home alone I saw a scroll hanging from the door knob. I picked it up and went to my bedroom. My 'orphan' wasn't there to read it to me; my 'orphaned' parents weren't there to listen to him, I had hardly spoken to them since the news of indictment. I unrolled the scroll and placed my looking glass in front of it.

Kepler finds the first laws of science in true sense, warned 'not to throw Christ's kingdom into confusion with his silly fancies', ...

'Nothing holds me; I want to give in to the sacred fury, I want to taunt mortal men with my candid confession: I have stolen the golden vessels of the Egyptians to make one of them a tabernacle for my God. If you forgive me, I shall rejoice. If you are angry, I shall bear it. The die is cast; I'm writing the book, to be read either now or by posterity, I care not which. It may wait a hundred years for a reader, since God has also waited six thousand years for a witness.'

Johannes Kepler (1571–1630), the German astronomer – small, wiry, sickly, impoverished, persecuted and unfortunate in his marriage – was wild with joy and mad with the excitement when on 15 May 1618 he discovered the relationship between a planet's period, the time it takes to complete a single orbit, and its distance from the sun: the ratio of the squares of the periods to the cubes of distances is the same for all planets. Known as Kepler's third law, it was published in 1619 in his book *Harmony of the Worlds*.

He had published his first two laws ten years earlier in his book *New Astronomy*: (1) the planets move around the sun in elliptical orbits with the sun somewhat off centre; (2) an orbiting planet sweeps over equal areas in equal times (this means the speed of a planet is not uniform; it speeds up when it approaches the sun and slows down when it's farthest from the sun).

These three laws provided a mathematical framework for the Copernican system and put an end to the obsession with Aristotle's perfect circles and the nightmare of Ptolemy's cycles and epicycles, explanation of planetary paths by circles on circles. Planets were now freely floating in the space, moved by physical forces acting on them.

'At last angels were not pushing the planets,' Chara blurts out. 'Industrious angels did a good job for centuries, didn't they, Thales?'

'Indeed, Chara.'

Science was at last divorced from religion and the teachings of ancients Greeks. Aristotle's cosmos started fading from the consciousness of scientists, but not religious leaders.

Copernicus' *Revolutions* had been placed on Catholic Church's Index of Prohibited Books. Kepler's passionate faith in the

Copernican system ('The sun not only stands in the centre of the universe, but is its moving spirit', he asserted) brought him the disfavour of religious leaders, and the title of 'mad stargazer' from the people. He was warned by the Church 'not to throw Christ's kingdom into confusion with his silly fancies' and ordered to 'bring his theory of the world into harmony with the scriptures'.

Kepler's laws are a major landmark in science; the laws were the first serious attempt to interpret the mechanism of the solar system. They were the first laws of science in the true sense: precise verifiable statements about natural phenomena expressed in mathematical terms. As a scientific law predicts the behaviour of a natural phenomenon, Kepler's laws could be used to predict the position of a planet at a given time.

Kepler was a versatile genius who, besides discovering these three laws, compiled the tables of star positions, improvised the telescope, observed a supernova and noted that it didn't show any parallax. When he turned his gaze from the heavens to the earth he worked on infinitesimal calculus and logarithms, founded the science of geometrical optics, examined snowflakes and described them in a fifteen-page essay that is the first book on crystallography, studied the anatomy of the human eye, explained the tides of the oceans and dreamed of a journey to the moon.

Parallax? I didn't know what it was. Cosmo wasn't there to explain it to me. I had to read a book. It said that parallax is the illusion of an object moving when seen from two different places.

It's damn boring reading a textbook. I wished I could summon the spirit of Thales. 'Oh, great Thales, what is parallax and what did Kepler conclude when he measured that the new star didn't show any parallax?'

I heard a voice of an old man. 'Jane, hold your hand out with thumb raised in front of any distant object. Now close each eye in turn and look at the thumb. The thumb will appear to move relative to the object. The more distant is the object, the smaller is its parallax.'

'Umm.'

The voice continued. 'When in 1604 Kepler observed the nova, he discovered that it didn't show any parallax. There were two possible reasons for this: either the earth was motionless at the centre of the universe or the stars were so far away that their parallax was too small to measure. He concluded that the nova lay within the sphere of distant stars and like the nova other stars were also changeable. This conclusion was at odd with the church's belief that the universe was unchangeable. His observation proved that the universe does change.'

'Umm.'

'When in 1572 Tycho observed a nova he also correctly noted that it didn't show any parallax, and also arrived at the same two possibilities. However, he made the wrong conclusion when he said that the earth was at the centre of the universe. This wrong conclusion also led him to reject the Copernican system.'

'Umm.'

'Jane, it's possible that like Tycho your father could also arrive at wrong conclusion from his observations of Mr Donne's death and the girls' seizures: either they were caused naturally or caused by witchcraft.'

Was I cracking up under pressure or was it a dream?

... in a dream he skilfully meshes fiction with lunar astronomy, and writes the first work of science fiction in the modern sense, ...

Kepler was a 22-year-old student at the University of Tübingen in Germany when, in 1593, he dreamed of writing a book: not an astronomical treatise but a science fiction. The idea occurred to him when he decided to devote his graduate thesis to the question: How would the heavenly phenomena appear to an observer on the moon?

This question was inspired by Kepler's enthusiasm for the new astronomy espoused by Copernicus, the man whose prophet he was to become. Twelve years passed before he wrote the first draft, and another twelve before he found time to look at it again. This provocative and innovative book, written in Latin, was published in 1634, four years after his sudden death.

Somnium (The Dream) describes a young boy's trip from the earth (named Volva in the book) to the moon (Lavinia). The boy, Duracotus, lives with his mother Fiolxhilde on an island. Fiolxhilda is a witch who summons up a demon to propel him to the moon. Kepler is somewhat vague about the method of propulsion, but his estimate of the distance of the moon is correct.

Unlike the ancient Greeks, Kepler knew that the earth's atmosphere does not extend as far as the moon. To enable Duracotus to pass through the thinning atmosphere, he is put to sleep with the aid of opiates and his nostrils stopped with moistened sponges. So that the boy's body is not torn apart by the great acceleration needed to escape the earth, he sits with his arms and legs curled inwards. As he is pulled by the demon towards the moon, he reaches a point 'where the moon's magnetic force balances that of the earth'. When the demon releases him, he falls to the moon unaided.

Kepler was highly influenced by Gilbert's *De Magnete*. While Gilbert thought that magnetism holds the world together and was the actual cause of the earth's rotation, Kepler went further and envisioned universal forces related to magnetism for controlling the planets. Thus he failed to see the true identity of the force which keeps planets in their orbits. A later scientist would reveal its identity and would call it gravity.

'Like spiders they will stretch out and contract, and propel themselves forward by their own force – for, as the magnetic forces of the earth and the moon both attract the body and hold it suspended, the effect is as if neither of them was attracting it – so that in the end its mass will by itself turn towards the moon.' That's how he describes the effect of space travel on his hero's body.

When his hero's journey is completed, Kepler proceeds to describe in detail the astronomy, landscape, climate and plant and animal life of the moon. The story ends when Duracotus is awakened from his dream by a cloudburst.

I wondered why a Renaissance scientist conjured up a witch to propel his fictional ship to the moon. He could have dreamed up a better way than relying on the powers of an old hag riding a broomstick.

... and vigorously defends his mother against the charge of witchcraft by presenting medical reasons for her so-called victims' illnesses.

Kepler's mother Katharina was well known in her hometown Leonberg as a quarrelsome old woman with vile a temper and a sharp tongue. She was brought up by an aunt who was burned at stake for witchcraft. Katharina dabbled in herbs and healing. When the herbs worked she was a heroine, when they didn't she was a villain.

She was in a habit of offering drinks to her guests in a tin jug. Once she gave Ursula Reinbold, a dried-up old woman, a potion in her tin jug, Ursula tasted and immediately shouted, 'Good Devil, what is this? What did you give me to drink? It is as bitter as gall.' Katharina ignored her complaint because she knew what a cantankerous woman her friend was.

In 1617 Ursula accused Katharina of giving her witches' potion which had produced a chronic illness. Various citizens of Leonberg suddenly remembered that they had also been taken ill by drinking from Katharina's jug. The butcher swore that her wife died of it; the schoolmaster alleged that he was permanently paralysed of it; a girl of twelve insinuated that when she saw Katharina she felt sudden pains in her arm and the arm had been temporarily paralysed.

As a result of these claims Katharina was charged with forty-nine counts of witchcraft, including that the accused had failed to shed tears when admonished with the text from the Holy Book. When this charge was read to her, she retorted that she had shed so many tears in her life that she had none left.

The proceedings continued for fourteen months, while 74-year-old Katharina spent her nights and days in chains in a room guarded by two guards whose salary had to be paid by her. Kepler, who was then working on *Harmony of the World*, mounted a thorough and vigorous defence. He showed that the butcher's wife died of abortion, the schoolmaster was paralysed when jumping a ditch and the girl's arm was temporarily paralysed when she was carrying bricks to a kiln.

The court observed that circumstantial evidence was not sufficient for the use of torture but also did not permit her release, and ruled that she should be questioned under threat of torture, not actually tortured. She was led into the torture chamber and the executor showed the rack, the wheel, hot poker and such other instruments of torture and explained their use in detail. This was her last chance to confess. She refused to confess, fell on her knees and prayed. 'I know God will bring the truth to light and will not take His Holy Ghost from me.'

She was sent back to prison. A week later the Duke of Württemberg ordered that she be released since she had cleansed herself through her ordeal. She could not return to Leonberg because of threat of lynching by the populace. She died six months later in 1622 from illnesses directly attributed to the rigours of her imprisonment.

So Fiolxhilde the witch was an image of his mother Katharina. Witch mania was not unique to our town, I worried.

My somnium was that if the doctor in the house failed to prove to the court the medical reasons for Mr Donne's death and five girls' craziness, then Dr Jane Digby, without a medical degree a quack at best, should face the judges and prove convincingly that disease and death are caused by natural causes not witchcraft. And then I should summon the spirit of Thales to tell the court that scientific method is the best tool ever devised to separate reality from humbug. Wouldn't believing in spirits be itself humbug, in the eyes of science?

~~~

It was the kind of day the cuckoo likes, bright and breezy. While she would be singing in the woods, our streets would be buzzing with rumours. Title-tattle would keep the citizens tireless on the day of the trial.

Cosmo and I stepped out of the house into golden sunshine which had been eluding us like mirage, ready to witness the drama in the courtroom, a drama better played in a theatre. 'Here's another scroll,' Cosmo whooped and picked it up from the door knob; since his mother's death I had been concerned about his mood swings. We kept walking towards High Street. 'It's bigger than earlier scrolls. Why is Thales burning the midnight oil? After reading I shall give it to Mr Johnson. He likes reading them.'

Our High Street was not really high in any sense. It was a small street with a row of shops. Among a baker, a butcher, a grocer, a shoemaker, a tailor, a draper and such shops of necessities there was an assembly room. It had been turned into a shop of justice to be run by the assizes judges.

'He says that Sir William Pelham would be hearing the case,' Cosmo said after a few moments. 'He reckons Sir William is a witch-monger and a bully; never acquitted a woman charged with witchcraft. The jury is always an ignorant parcel of men. So he will have his way.'

'Don't bury the poor woman before she is dead.'

'I wish Mr Johnson would be selected as a juryman. His name is on the list.'

'How would it help?'

'The trial jury is the jury of life and death. Its decision had to be unanimous and Mr Johnson will never cave in.'

We were in High Street. I had never seen so many people, peddlers, horses, coaches, drays in our quiet street. In ye olde moo moo world we made the best of any opportunity to entertain ourselves.

The public gallery for spectators overlooking the courtroom was already packed. There were no benches for spectators on the ground floor, only standing room. We stood, as most of us did in theatre, to watch the show. The judge's bench was empty, but there was a man at clerk's desk, with inkwell filled and quill pen sharpened to his contentment, ready to scribble on the stack of creamy white paper in front of him. Mrs Marsh was in the dock, a chain dangling from her feet, Broderick and Bowes stood on either side. She looked alert; Broderick had obviously cut down her daily supply of ale. I could see Salus and Mama chatting with the parents of the famous five in the front row of the benches reserved for witnesses and town dignitaries. The actresses, er girls, were muted, but Vicar standing near the bench was animated, his body language apparently apprising his parishioners that though there was a Satan's agent in the courtroom, God was also fully represented.

A middle-aged man wearing a silk-lined scarlet robe and a black cravat entered the room from a door near the bench; the sides of his judicial white face were obscured by a white shoulder-length horsehair wig. They say wigs were scented with pomade. I took a deep whiff but could only smell of unwashed bodies around me. 'He's Judge Pelham,' Cosmo whispered, 'whosoever said the judges look like mice peeping out of oakum was right.' The white mouse sat in a plush chair behind the bench, everyone stood up.

The clerk with his white staff in his hand nodded to the town crier. He cried. 'The King's Judge straightly charge and command all manner of persons to keep silence, and hear him upon pain of imprisonment.' The clerk said 'God Save the King' and the crier repeated the same after him and then a man sounded the trumpet.

The proceedings of the court had now begun, but there was no silence. Like our theatres, our courts of law were rowdy places. Both did put on shows; to the public it didn't matter whether the show was about Macbeth or Mother Marsh.

The clerk passed on a piece of paper to the crier. Cosmo breathed heavily. 'It's the list of jurymen,' he whispered. The crier shouted. 'You good men that be returned to enquire before for our Sovereign Lord the King answer to your names every man at first call, and save your fines.'

Under the threat of fine twelve men said 'Aye' when their name was called, 'Master Gifford, Master Landish, Master (scarecrow), Master (hollowed-cheek), Master (gangling-cadaver), Master (beak-nose), Master (middling-height), Master (thin-reed), Master (bull-necked), Master (cross-eyed), Master Sampson, Master Johnson'. Hearing the last name, Cosmo squealed like a mouse. Mouse at the bench glanced at him but ignored him when he saw another male. I turned my face away as I was in no mood of mating with His Mouseship.

The bailiff stood up with the Bible in his hand. He went near the dock and babbled in one breath. 'Mother Marsh, place your right hand on the Book you shall truly answer to such questions as the Court shall demand of you, so help you God.' Mrs placed her hand on the Book. The bailiff's babbling brook was now flowing downhill and had picked up speed. 'These good men that were last called and do now appear are those that shall pass between our Sovereign Lord and the King and you upon your life and death, if therefore you challenge any of them, you may challenge them as they come to the Book to be sworn.'

Mrs Marsh said nothing. There was no Cicero to defend her; our lawyers were for the rich and famous.

After swearing in the jury, the bailiff read the charges.

'Are you guilty of these charges,' His Mouseship roared.

'Not guilty,' Mrs Marsh whimpered.

'How will you be tried?'

Mrs Marsh was expected to say, 'By God and the Country.'

Mouse didn't wait for her reply. 'God send you a good deliverance.'

Lulled by Mr Johnson's presence in the jury box, Cosmo's was more interested in the words apparently penned by the godly spirit of a scientist than the words being spoken in the courtroom in the trial of a devilish woman.

~~~

Tried for heresy, the great scientist confesses under the threat of torture, placed under the house arrest for the rest of his life, his book banned, ...

22 June 1633. Ten stern-faced cardinals of the Inquisition at Rome, each splendidly dressed in a scarlet cassock, white surplice trimmed

with lace and a short hooded cape fastened with pairs of button down the front, are sitting on finely carved chairs behind a large table in the forbidding Hall of Inquisition in Rome. A frail, old man dressed in a white gown of penitence is kneeling at the centre of the Hall. He speaks in a soft, subdued voice.

'I, Galileo Galilei, son of the late Vincenzio Galilei of Florence, aged seventy years, being brought personally to judgment ... having before my eyes and touching with my hands the Holy Gospels, swear that I have always believed, do believe and by God's help will in the future believe all that is held, preached and taught by the Holy Catholic Church ... I must altogether abandon the false opinion that the sun is the centre of the world and immovable and that the earth is not the centre of the world and moves and that I must not hold, defend or teach in any way whatsoever, verbally or in writing, the said false doctrine ...'

Legend has it that the great scientist was so convinced that it is the earth that moves, as he rose from his knees stamped his foot on the ground and muttered under his breath, *'Eppur si muove'* ('And yet it moves').

With this public recantation ended the most tragic trial in the history of science. Though saved from the Inquisition's deadly dungeons when the Pope commuted his sentence, he was placed under house arrest for the rest of his life; his book was banned and placed on the Index of Prohibited Books.

Galileo was the first to study the heavens through the telescope, which he invented independently in 1609 'by means of reasoning' when he heard the news of a spyglass made by a Dutch spectacle maker Hans Lipperhey. Everything he saw through his telescope convinced him that Copernicus was right. The earth and the planets not only spin on their axes; they also revolve about the sun in circular orbits. Dark 'spots' on the surface of the sun appear to move; therefore, the sun must also rotate, he concluded. The craters and mountains on the moon and the sunspots, he argued, prove that the ancient theory of perfect heavenly spheres was wrong. Like Bruno, he also believed that indeed there might be life on other planets. As a scientist he was interested in the possibility of life on other worlds, he said, and not interested to venture into the religious question whether or not Christ had redeemed these extraterrestrial creatures.

He didn't dare to publish his views 'fearing,' he wrote to Kepler, 'the fate of our master, Copernicus.' Kepler, his comrade in the pursuit of truth, replied, 'It is not only Italians who cannot believe that they move if they do not feel it, but in Germany also one finds little favour with this idea. Be of good cheer, Galileo, and come out publicly ... great is the power of truth.'

In 1632 Galileo published his masterpiece, *Dialogue Concerning the Two Chief World Systems*, in which he eloquently defended and extended the Copernican system. Against the universal practice of his time to write on philosophical and scientific subjects in Latin, Galileo wrote in Italian 'because I wished everyone to be able to read what I wrote.' His views were thought to contravene the teachings of the Church. A year later, he was tried for heresy by the

Inquisition and forced to renounce his theories. Under the threat of torture he agreed to 'confess', against his belief that 'the Bible teaches us to go to heaven, not how the heavens go.'

In this book, which marked the transition from the dark days of the Middle Ages to the beginning of the era of modern science, the dialogue is between Salviati, a philosopher who represents the views of Galileo; Simplicio, a pedantic who is a follower of Aristotle; and Sagredo, an intelligent layman.

Simplicio: How do you deduce that it is not the earth, but the sun, which is at the centre of the revolutions of planets?

Salviati: Neither Aristotle nor you can prove that the earth is de facto centre of the universe; if any centre may be assigned to the universe, we shall rather find the sun to be placed there. This is deducted from the most obvious and therefore most powerfully convincing observations.

Simplicio: Please, Salviati, speak more respectfully of Aristotle.

Salviati: Does not he say that a circular motion for the earth would be forced, and therefore not eternal? And that this is absurd, since the world order is eternal.

Sagredo: What kinds of things are the earth, the sun and the stars in nature? Are they trifling things or important?

Simplicio: They are principal bodies; most noble, integral parts of the universe.

I nudged Cosmo and whispered. 'Noble bodies are not in the heavens, they are right in front of you.' 'Good heavens!' Cosmo replied rather loudly, 'I can't see any noble bodies at the bench, only a mouse.' Laughter shook the courtroom. I didn't laugh, but felt like tickling the cuddly mouse at the bench. Why one cannot tickle oneself? Before laughter could shake the skies, the mouse mumbled 'Silence!' to extinguish it. The bailiff shouted, 'Silence!' The crier bellowed, 'Silence!'

The clerk stamped his white staff on the ground. The bailiff barked, 'Jeffrey Gooding to take stand.' Even before Mr Gooding could reach the witness stand, the bailiff started babbling. 'The evidence that you shall give to this inquest against Elizabeth Marsh, prisoner at the Bar, shall be the truth, the whole truth, and nothing but truth. So help you God', and pushed the Bible against Mr Gooding's hands.

His Mouseship looked at the witness expectantly, Mr Gooding obliged. 'My Lord, last summer the axle of my cart, happening, in passing to break a broom leaning on the fence of the witch's cottage.'

'The witness to address the accused as Mother Marsh,' His Mouseship warned.

Ignoring him, Mr Gooding continued. 'She became very angry and threatened me that my horses would suffer for it. After this incident all my four horses died and I suffered sharp pains in my both knees. Dr Digby examined me thoroughly and was surprised how fast the condition had become worse. Then my head was filled with lice. They were so many and so big that I couldn't sleep at night. Goodwife Bowes advised me to burn two pairs of clothes similar to the one I was wearing at the time the lice swarmed to undo the witch's curse.'

His Mouseship seemed interested in this new but expensive remedy for witchcraft. 'Did the burning of clothes helped?'

'Yes, my Lord.'

Mr Johnson raised his hand. 'Permission to question the witness, Your Lordship.' His Mouseship nodded; the jurymen were allowed to examine witnesses.

'Master Gooding, how long after the accident did the horses die?'

'After a short time.'

'How short?'

'It was after the feast of St Mary Magdalene.' He counted on his fingers. 'Two months, may be three.'

'Wouldn't witches' spells work much faster?'

Gangling-Cadaver spoke in a voice as if it was from the other world. 'You're mistaken, sir, it could even take years.'

Mr Johnson turned towards the bench. 'Your Lordship, do I have your permission to question Dr Digby.'

His Mouseship shook his head. 'No, the jurymen aren't allowed to question members of the public. You may question Dr Digby when he takes the stand.'

Pleased with Mr Johnson's cross-examining skills, Cosmo was again in Italy.

... checks his pulse and discovers that a swinging pendulum can be used for timekeeping, ...

In 1581, when Galileo was a seventeen-year-old medical student at the University of Pisa, he made a startling discovery: A pendulum will swing at a constant time, which means it can be used for timekeeping.

During Mass at the Cathedral of Pisa, he became bored and dreamily fixed his eyes on a chandelier swinging from a long rope. It seemed to him that the time of the swing was the same whether the swing was a large or small one. He used his own pulse beat – as a medical student he knew that under normal conditions our pulse beats regularly – to test his intuition.

Later on he experimented with a metal ball suspended by a string – what's now known as a simple pendulum – and found that he was correct. Every swing of the ball, large or small, took the same time. This is called isochronism of the pendulum.

In 1602 he used the principle of pendulum to invent an instrument to measure the pulse rates of patients. The simple device, the pulsilogium, proved of great value to physicians. Years later, in 1641, at the age of seventy-seven when he was totally blind, the idea of making a clock regulated by a pendulum occurred to him. His son, Vincenzio, a clever mechanic, made several drawings and models. However, the first working pendulum clock was made by the Dutch scientist Christiaan Huygens in 1656.

Galileo's discovery of the principle of pendulum paved the way for the accurate measurement of small intervals of time.

Mrs Pollard was now on the witness stand. I wished I had Galileo's pulsilogium to check her pulse rate as she was nervous and agitated. His Mouseship whispered something in the ears of his clerk. The clerk took a chair and asked her to sit down. 'That witch,' she pointed at Mrs Marsh, 'had turned my two lovely daughters into little monsters. My girls would see mice running round the house, catch them and throw them into fire, they would screech out like rats.'

'Did you see the mice?' His Mouseship enquired in a sympathetic voice. 'How big they were?'

'No, my Lord, I never saw any mice. There are no pests in my house. I don't even allow cats and dogs inside.'

'Did you hear the screeching?'

'Yes, my Lord, quite clearly. One day Grace caught a mouse and threw it in fire, it flashed like gunpowder. Bang! Bang! Bang! No one saw the mouse, but everyone saw the flash. You can ask my husband, my two maids. They all were there, my Lord.'

First time since the court proceedings had started there was complete silence in the courtroom.

His Mouseship looked at Mrs Pollard with sympathy.

'Once something like a fly flew at Joan's face. She vomited up a two-penny nail.' She started sobbing. 'I'm sure that witch had sent the fly with the nail. It forced into my sweet daughter's mouth.'

She slumped into chair and then fell on the ground. Salus came running towards her.

... drops two lead balls from the Leaning Tower of Pisa and shatters ancient views on the motion of falling bodies.

Aristotle was the first to speculate on the motion of falling bodies. Without any experimental evidence he concluded that the heavier the body was faster it fell, and the thicker the stuff it fell through, the slower it fell (for example, a stone will faster in air than in water).

It was not until eighteen centuries later that this notion was challenged by Galileo. He designed an inclined plane that had a small groove cut along its entire length. The height of the plane could be varied and the groove was covered with a smooth parchment to reduce friction. He conducted meticulous experiments with brass balls on his inclined plane to study the motion of falling bodies. From these experiments he formulated his laws of falling bodies. The laws were published in *Dialogue*: Discounting air resistance, all bodies fall with the same motion; started together, they fall together. The motion is one with constant acceleration: the body gains speed at a steady rate; that is, it gains the same addition of speed in each successive second.

Legend has it that in 1591 Galileo, then a young professor at the University of Pisa, climbed the spiral staircase of the leaning tower carrying with him two lead balls, one weighing one hundred times more than the other. When he reached the gallery surmounting the seventh tier of arches, he carefully balanced the balls on the edge of the parapet and rolled them over together. Students and staff of the university assembled near the foot of the tower were amazed to see the two balls, falling together and reaching the ground at the same moment, when they expected the heavier ball to fall hundred times faster. The loud clang of the balls striking the ground together once and for all shattered Aristotle's views on the motion of falling bodies.

One of the last major figures of the Renaissance, Galileo was born in 1564, a year that saw the birth of Shakespeare and the death of Michelangelo. It's tragic that he was caught in the conflict between a retreating age of belief and an advancing age of reason. When he died in 1642 as a martyr of thought his fight for the freedom of belief had not been truly won. The ideal of his declaration 'Philosophy wants to be free' was still shackled, but Galileo – and Kepler – had made those shackles brittle and fragile.

Cosmo and Jane, I wonder whether humans will ever truly win the fight for freedom of belief.

'At least, our reverend has truly won the fight for freedom of his belief,' Cosmo said sarcastically when he heard the bailiff calling Vicar to the stand.

'God denounces those who conjure spells and those who practise witchcraft and sorcery.' Vicar began as if he was delivering another sermon. 'Only God has the right to understand the realm of the supernatural. The evil Queen Jezebel practised witchcraft brining catastrophe on herself and all Israel.'

'You're absolutely right Vicar,' Hollowed-Cheek interrupted. 'Intrusion in the realm of occult makes one worthy of death.'

Scarecrow opened his Bible, showed a page to the public. 'It's written here that their place will be in the fiery lake of burning sulphur.'

Not to be the one left behind, Mr Johnson also forgot his well-honed courtroom etiquettes, stood up and shouted at his fellow jurymen. 'Why shall we imagine that this poor doting old woman is like any of the witches mentioned in the Holy Book? We know nothing about the nature of spirits and their power of acting upon humans.'

His Mouseship was taken aback by the uproar in the jury box. He said in his loudest classy voice, 'Order!' The bailiff shouted, 'Order!' The crier bellowed, 'Order!'

'Reverend, please tell the court what makes you believe that the accused is a witch.'

Vicar smiled. 'Yes, My Lordship. First, this wicked woman couldn't recite the Lord's Prayer. Have you ever heard of a Christian who refuses to say the Lord's Prayer? It's the Devil who controls her is refusing to praise our Lord. Second, bewitched children are usually cured when they scratch the witch. Her pact with the Devil is so strong that five girls are still suffering from terrible seizures and spells after scratching her.'

Cosmo was probably not interested in hearing about the Devil's pact with Mrs Marsh; he unrolled the scroll and said:

Ah, here's a man who thinks before he acts. And when he thinks he redraws philosophy, physics and mathematics.

In his early years the French philosopher and mathematician René Descartes (1596–1650), a contemporary of Galileo, was sceptical of almost everything, even his own existence. He lost this scepticism after reaching the conclusion, *'Cogito, ergo sum'* ('I think, therefore I am'), philosophy's most famous statement.

When he was sent to a boarding school at the age of eight, he enjoyed exceptional privileges because of his poor health. 'My philosopher', as his father used to call him, was excused from morning school duties and was allowed to stay in bed until late in the morning. This habit of morning reflections in bed clung to him throughout his life.

In his later life, Descartes, a small timid man with large head, low forehead and the discreet, stubborn and fanciful eyes who spoke in a feeble voice, reflected upon how to arrive at knowledge without any fear of error. He concluded that only two mental acts, intuition and deduction, are the two most certain paths to knowledge. Intuition, he said, is the undoubting conception of a pure and attentive mind. Intuition proceeds by deduction, by which we understand all that is necessarily concluded from certain other facts already known.

To discover what, if anything, he can know with certainty, he presents a series of arguments to cast doubt on knowledge he has accepted as truth: All those things that have entered my mind were no truer than the illusion of my dreams. Or, a deceiving God or an evil demon is causing me to go wrong about knowledge, which is self-evident and which I seem to see so clearly. Because my senses may sometimes deceive me, the source of my knowledge cannot lie in my senses. I can only be certain about the experience of thinking. I cannot doubt that I think, therefore, I exist as long as I'm thinking.

After being assured of his existence, Descartes wonders if his existence implies the existence of God. You cannot even think of the idea of God without thinking He exists. Thus, by definition, God exists. God is perfect and therefore He doesn't deceive us; now I can trust my ideas and can build a perfect mathematical system of knowledge.

The child of this philosophy was Descartes' physics. Like his predecessors, he did not criticise Aristotle's physics, but created an entirely new physics to explain the nature of the universe: Both matter and energy were conserved from God's original creation and the universe was derived and maintained by general laws ordained by God. Everything in nature could ultimately be reduced to the rearrangement of particles moving according to these laws. Space by itself being nothing, it has no extension – no length, breadth or height. Only matter has the property of extension, and space cannot exist where there is no matter. Matter exists everywhere, and a vacuum exists nowhere. However, his laws of nature required more time to build the universe than the six days taken by God, according to the Bible, to create the sun, earth, moon and mankind.

Descartes was thirty-seven when he presented his theory in an imposing book, *The World*. The book endorsed the sun-centred universe of Copernicus and Galileo and he feared that it would offend the Inquisition at Rome. When he was giving finishing touches to his book, he heard the stunning news of the trial of Galileo. The news crushed him. He was as convinced of the truth of the Copernican system as he was of his own existence. But he was also convinced of the great power wielded by the Inquisition. As he had no desire to become a martyr, he decided not to publish his book. A decade later, in 1644, he published his most comprehensive work on physics, *Principles of Philosophy*, which was an extension of *The World*. The Church immediately placed it on its Index of Prohibited Books.

After defining space, Descartes turned his thinking to defining his position in space. A point in space can be completely fixed if we know its distances from three arbitrarily chosen lines of references that are at right angles to each other. This idea gave birth to a new geometry, now known as analytical geometry, which uses methods of algebra to solve the problems of geometry. In this system, the lines of reference are usually referred to as x-, y- and z-axes, and distances from these lines are labelled x, y and z and are called the coordinates of the point.

'You know Thales, Descartes came up with the idea of analytical geometry,' Chara says, 'when he watched a fly crawling on the ceiling of his room, and realised its position could be defined by its distances from the two adjacent walls.' She is lying on her back and looking at the ceiling hoping to discover another way to define her position in the four-dimensional universe, three dimensions of space and the fourth dimension of time.

She tweets, 'Thales, you're the Father of Western Philosophy. Am I right to call Descartes the Father of Modern Western Philosophy?' She does know Descartes' and my true place in history.

The bailiff called Mary Legat, a barmaid at the Bull. The nervous young thing was so awestruck by the court goings-on that she started burbling her statement even before the bailiff had pushed the Bible against her hand. 'My Lord, Mr Donne was a very kind man. A perfect gentleman, sir. I can never forget the day when I saw him disappearing right before my eyes without a trace.'

His Mouseship looked at her deposition. 'After Mr Donne went into flames there was nothing left, no ashes, absolutely nothing, are you sure?'

'Yes, my Lord. No smell either.'

'Did Mr Donne tell you about being bewitched?'

'Yes, my Lord, the morning he died he told me that the night before he couldn't sleep. Someone was beating drums in his house all night. He got up and looked around and saw nobody. I know who the drummer was?'

His Mouseship looked at the deposition again. 'Nothing here about any drummer.'

'Constable Bowes must have overlooked it. I can't blame him. He took so many depositions on that day. He really worked hard on my deposition.'

Sure he did, I thought. Our dopey constable was more interested in flirting with fillies, real or imagined, than ferreting facts. Even now he was ogling at the barmaid's ample bosom, sizing them up with those in saucy stories.

After portraying PC Plod as PC Pro, the flustered filly became quiet. His Mouseship's face gave a hint of smile. 'You need not worry about what's in the deposition. You are allowed to tell the court now.'

The small dose of judicial smile put her at ease. 'A few months ago a drummer who lived near our house died. They say that one evening when he was walking past the witch's house, he was suddenly lifted in the air and then hit the ground. I'm sure it's his ghost that was playing drums in Mr Donne's house.'

His Mouseship craned his neck, eager to add another witchcraft tale to his repository.

'On the morning of Mr Donne's death, I saw a cat near the stool he was sitting on. Mr Bartlett doesn't like cats or dogs coming into his barroom, so I shooed her away. Mr Donne asked, what are you doing, Mary? I replied, Mr Donne, there is a cat near your feet. He said I can't see no cat. Within seconds he was in terrible pain.'

Bull-Necked stood up, snapped his finger and then pointed it at Mrs Marsh. 'That has to be her cat. My Lord, a bewitched person can't see the witch's cat. Everyone here will agree with me.'

His Mouseship yawned; perhaps he had heard this tale before. The bailiff burbled. 'All manner of persons that have appeared here before my Lo rd the King's Judge may take their ease at present and attend here again at two of the clock in the afternoon, God Save the King.'

Sword

The clerk stood up with the white staff in his hand. A man sounded the trumpet. The court was in session again. Cosmo and I had spent the past two hours wandering aimlessly, chewing pieces of bread, cheese and corned beef I had brought with me in a cambric bag. No fancy picnic baskets for us in the church's garden; no lavish lunch either at the dandy's manor. Cosmo and I were distressed by the news that the doctor, the vicar and the judge were Cambridge cronies, the kind that manifests masculinity by 'best of luck, old chap' back-slapping.

I had also heard the rumour that the judge and his rather youngish wife were lodging at the Bull. I had seen the snooty thing riding around the town. Well, the Bull's famous bed was big enough to try out her riding skills on her cuddly white mouse. While his trophy wife was back in her vast and quiet room, His Mouseship's courtroom was crowded and noisy. More spectators had come in; seasoned ones knew when the courtroom staged the best dramatic scenes.

Broderick was on the witness stand. He was in his element; gaoler's testimony was valued in a witchcraft trial. 'I have found Mother Marsh very obstinate and unruly. One evening I had to chain both her legs. She pushed me so hard that I asked my wife for help. The witch grabbed my wife's hair and cursed her. That night my wife became terribly sick, her body heaving up and down, her arms, legs and head shaking.' He pointed at the girls. 'Just like those children, Your Lordship. I brought the witch to my wife's bedside and held her there until my wife had scratched her face. Within minutes my wife recovered completely.'

What my foolhardy friend had bought with his bribe of five gold coins? I wondered. A gallon of ale for the old woman and a wagon full of lies for the court?

Invigorated by the lunch at the manor, the judge was in an inquisitorial mood. 'Had your wife suffered such symptoms before?'

'No, Your Lordship, she is as healthy as a bull.'

His Mouseship chuckled. 'Have you been watching the prisoner for any signs of imps?'

'Yes, Your Lordship, imps visit her every night.' He placed his index fingers side by side. 'The bars on the cell's window are that close that even a tiny mouse couldn't squeeze through them, but I could hear a cat miaowing in her cell. It happened every night the moment I locked the door.'

His Mouseship nodded.

'One morning I asked her, Mother Marsh, last night I heard a man's voice in your cell. Was there someone with you? Yes, she replied, a man came to my bed and said, Bess, your sins are as red as scarlet, if you sleep with me I'll make them as white as snow.'

His Mouseship adjusted his wig and directed the clerk to note down every word of the testimony and then asked the witness. 'Did she tell you how the man was dressed?'

Pleased with the importance attached to his witness statement, Broderick spiced up his pot of lies. 'The man was wearing a black tunic, green stockings and a high black hat with a red feather on it. He had goat's feet.'

A stereotypical description of the Devil when he comes to visit a witch, I thought. A young man standing nearby whispered. 'Did you know what does the feather on the hat refer to? A sugary lass giggled without a trace of shyness. 'That little thing in your breeches.' A modern girl, not a modest girl like me.

The gaoler was still babbling. 'The witch then told me that the man demanded some of her blood. She cut her finger with a knife and the man caught the blood in his hand. He then dipped a sharpened stick into the blood, gave it to her and guided her to sign her name in his book.'

Scarecrow stood up, made the sign of the cross and blurted. 'God save us. This woman has made a diabolic pact with the Devil. She had slept with him.'

The courtroom became abnormally silent. Many in the jury box and the public gallery were on their knees, praying. As a woman who had abnormal ambitions of becoming a physician, I was familiar with exotic viruses. I said to Cosmo. 'The weak-knees virus is highly contagious. Should I hold you?'

Silence didn't last long. Someone from the public gallery shouted. 'Hang the witch, Judge.' This shout was echoed a hundred times. This prompted the crier to repeat the warning of hearing the judge upon pain of imprisonment.

The bailiff bellowed. 'Amy Bartlett, Agnes Bartlett, Alice Bartlett, Grace Pollard and Joan Pollard to take stand.'

As the girls lined up on the witness stand His Mouseship stared hard at Mrs Marsh. 'Mother Marsh, I warn you not to look at the children. You shall face the jury.' This was a precaution against the evil eyes of witches.

He then turned towards the girls and spoke softly. 'You would be sworn on the Holy Book and would be speaking under oath. Suppose you should tell a lie, do you know what will become of you?'

The girls just looked at the cuddly mouse admiringly.

His Mouseship replied for the girls. 'If you tell a lie, you will go to the hell. Do you know who is the father of lies?'

'The Devil,' Amy replied quickly. 'That witch sleeps with the Devil.' Agnes, Alice, Grace and Joan repeated loudly, 'That witch sleeps with the Devil.'

His Mouseship remained quiet. Obviously he was used to dealing with unruly and bewitched children in his court. 'Look upon the prisoner in the dock and tell me did you know her. Is she is the woman who had afflicted you?'

'Yes, she is the witch,' Amy said. The other girls shouted. 'Hang the witch, hang the witch, hang the witch.'

Mrs Marsh turned around. 'I have done no harm to these girls.'

The moment Mrs Marsh looked at the girls, they became hysterical, their bodies in convulsion.

'Here we go again.' Cosmo broke his long silence. 'I'm in no mood to watch the same drama again. Let's us go outside. I want to sit down for a while.'

We went outside and sat under an alcove. He showed me the scroll he had been carrying with him the whole morning. 'I haven't finished reading it. It's the last one from our dear Thales. I'm sorry, Jane, so far he has failed to provide any answers to my dilemma.'

Pascal plays dice and proves that God exists, and agrees that he cannot explain it by scientific reasoning.

Two games of dice popular were once popular in the casinos of Europe. In the first game, roll a die four times and win if a six would come; in the second game, roll two dice twenty-four times and win if double six would come. Which game you would put your shirt on? The Chevalier de Méré, a nobleman described by his friend the mathematician Blaise Pascal (1623–62) as a man having a very good mind but no mathematics, thought the chances of winning would be better in the second game. But when the keen gambler continued losing his money, he asked his friend: Why? Pascal worked with fellow mathematician Pierre de Fermat to apply reason to randomness. Their hard analysis gave birth to the theory of probability.

Probability deals with the chances of an event happening in an unpredictable way. By snatching numbers from fortune tellers and giving them to mathematicians it helps us to cope with uncertainty. We can calculate everything from our chances of winning a lottery to our chances of being struck by lightning.

We can find the probability of an event by simply dividing the number of ways the event can happen by the total number of possible outcomes. This rule can be applied to tossing coins, rolling dice, dealing cards or drawing lottery numbers. What is the probability of drawing an ace of hearts from a well-shuffled pack of cards? There are four aces in a pack of 52 playing cards, the probability of drawing an ace is 4/52 or 1/13 and the probability of drawing an ace of hearts is 1/52.

'What is the probability of the jury of twelve acquitting Mrs Marsh?' I asked.

'Absolute, as long it has one enlightened member such as Mr Johnson; zero, if, God forbid, something happens to him,' Cosmo replied.

'There're no mathematical calculations involved in your answer; it all depends on divine calculations. What's then the use of studying probability?'

'If you let me read further you'll learn how probability can make you believe in divinity.'

Pascal's contribution to mathematics goes beyond probability theory, but he is now more widely known for *Pensées* (Thoughts), a collection of meditations on the nature of human life.

He makes at least one application of the theory of probability in one of his meditations, now known as Pascal's Wager. God either is or He is not. If we bet on whether God's exists, there are two chances. If we win the bet the reward is enormous (eternal happiness) and the loss is insignificant (only the time we have to spend in worship). If we lose the bet, still we lose little. Believing in God is the more sensible wager.

If the agnostic remains unconvinced, Pascal offers a different way of looking at the odds. If there were infinite eternal happiness to be won by leading religious life, there is one chance of winning against a finite number of chances of losing. Your chances of winning or losing are finite. You are playing for even odds as there are as many chances on one side as on the other. Therefore, it will pay to lead a religious life.

Pascal said, 'the heart has reason, which reason cannot know'; therefore, our belief in God cannot be explained by science's reasoning.

I smirked. 'Well, you know how you can save yourself from eternal damnation?'

Cosmo's face brightened.

I giggled. 'A divine revelation?'

'Yes. The great mathematician has revealed that spirituality minus science equals ignorant orthodoxy, but spirituality plus science leads to enlightened knowledge.'

~~~

When we came back into the courtroom we didn't see the girls; perhaps the judge had ordered them to be removed from the courtroom. Salus was on the witness stand.

His Mouseship smiled and asked in a voice exclusively blessed by Cambridge gods. 'Dr Digby, could you please describe to the jury the symptoms of the girls you have been treating so diligently, sir.'

Salus smiled back and replied in his best Cambridge accent. 'Your Lordship, any good physician would describe them as strange seizures with grotesque body movements, paralysed limbs, difficulty in breathing, seeing, hearing and speaking. At times my patients have been in unusually quiet and in prolonged trances, at other times they have been screaming violently.'

'Have you seen similar seizures in other patients?' the judge inquired.

'Not exactly similar, Your Lordship. The girls seem to suffer from afflictions medicine cannot explain.'

'From your deposition I understand, sir, you have ruled out their afflictions as hysteria or epilepsy.'

'Yes. I have never seen anyone suffering from hysteria or epilepsy coughing nails or pins. I can't think of any plausible cause for the numerous scratches on the whole body of one of the young girls. After the fits the girls would be unusually calm and would remember nothing.'

'Dr Digby, you're a learned and wise physician highly esteemed by people not only in these parts but as far as London. The Royal College of Physicians values your expertise. Is the jury right to conclude from your medical observations that there is neither a natural cause nor a natural remedy for the girls' afflictions?'

The mention of the Royal College had brightened the doctor's face. 'That's my conclusion, Your Lordship.'

'You have supervised the examination of Mother Marsh for any signs of Devil's marks. Do you wholeheartedly support the findings of the women jury appointed by the justice of the peace?'

'Yes, My Lordship.'

Cosmo's face turned blue. I started sobbing.

His Mouseship gave a Cambridge smile to his crony. 'Thank you, Dr Digby. You may now leave, sir' which translated into my ears as 'Well done, old chap.'

Mr Johnson raised his hand. 'The witness cannot leave yet, Your Lordship. The jury has the right to question the witness.'

His Mouseship was highly annoyed by this interruption. 'I'm amazed, sir, a building contractor who has spent much time in courts doesn't know the difference between an expert witness and ordinary witness. Ordinary witnesses are sworn to tell the truth, what they have heard or seen but not what they believe. As an expert witness Dr Digby is allowed to give his opinion. Your objection is overruled.'

'Thank you for explaining finer points of law,' Mr Johnson replied in a highly sarcastic and hostile tone, 'which I've forgotten in my old age. Do I have your permission to question Master James Dorrington?'

'Who is this Dorrington? He is not on the witness stand. You can't question a ghost, sir.'

'This ghost has submitted a deposition to the justice of the peace and as a member of the jury I'm allowed to question him. Should I point out where it's written in the assizes' proceedings?' He showed a book. 'If you haven't you can read it now. I have the latest sixteen-eighty-two edition here.'

The clerk searched through his pile of papers, pulled out a paper and put it on the bench. His Mouseship's ruddy face had turned ruddier. He nodded to the bailiff who barked. 'Master James Dorrington to take stand.'

His Mouseship started reading the paper and after a minute raised his head and asked Cosmo in a tone practised by the landed gentry on their serfs. 'This deposition has been written in a way suggesting that you dabble in medicine. What are you? A quack? A faith healer?'

'I'm an apprentice to Dr Digby, Your Mouseship.'

I was shocked at Como's slip of the tongue. Luckily, no one heard him correctly. 'Apprentices are not allowed to speak without their master's permission.' He turned towards Salus. 'Dr Digby, do you give permission to your apprentice to speak to the court? You have the right to refuse permission, sir.'

Salus replied. 'James is an intelligent lad and shows great promise of becoming a good physician. I have full faith in him.'

His Mouseship looked puzzled. 'Oh, you master speaks highly of you.' He turned towards the jury box. 'Mr Johnson, you may question the witness.'

Mr Johnson immediately obliged. 'Master Dorrington, please tell the court what you have submitted in your deposition.'

In my eyes the young man on the witness stand was as an incarnation of Paracelsus. He looked at the jurymen straight into their eyes and spoke in a measured voice like a prophet. 'If we cannot find medical causes for the girls' illness, we can come up with many other hypotheses. One of them would be that they have been bewitched. We will have to reject this hypothesis not only because there is no scientific evidence for witchcraft but the Holy Book doesn't teach us to believe in witchcraft.'

The Scarecrow stood up. 'It's blasphemy, Your Lordship.'

His Mouseship immediately shut him up. 'Let the witness finish his statement.'

Cosmo continued. 'The other hypothesis is that the girls are suffering from delusion. But again, many of you would say that this delusion has been caused by witchcraft.'

The Scarecrow stood up again. 'You're absolutely right here.'

His Mouseship stared at Scarecrow. 'If you interrupt again, sir, I'll be forced to remove you from the jury box.'

Scarecrow waved his Bible. 'You can remove me, but you cannot remove this from your courtroom, Your Lordship. What's written here is the truth.'

His Mouseship looked at Cosmo. 'Go ahead, speak the truth.'

'The third hypothesis, which I consider the most plausible, is that when the Bartletts came to these parts from Cambridge, their daughters were bored, especially the oldest one. The Bartletts are old fashioned and had placed many restrictions on their daughters. When the girls started acting as if they were bewitched by their neighbour, a sweet old lady, they received special attention and love and were exempt from punishment. Their first little play was soon out of their hands. It took on a life of its own.'

His Moughship interrupted. 'What I saw here this morning was five children suffering from an evil hand. Young girls screaming, crying, choking and convulsing distressed me and their parents. You're insulting this court and citizens of this town when you suggest that the girls are merely playing a game. You fancy yourself as a bard, don't you? This court is not a stage.' He waved his finger towards his bailiff. 'Remove this bard from my courtroom.'

As Cosmo started walking towards the door, His Mouseship said. 'Bailiff, you may allow to stay him in. I'll excuse him as a favour to his master.'

Cosmo joined me in the back of the courtroom. The sugary, flirty lass placed her hand on his shoulder. 'Well, done, sir, rich people don't like new ideas.' I gave her an evil eye and pushed her hand away from my Cosmo's shoulder.

The bailiff babbled his well-practised oath, took Mrs Marsh's hand and harshly pulled her from the dock to the witness box.

His Mouseship said. 'Elizabeth Marsh, did you not see what you did to five little girls in my court? Why do you such torment these poor girls?'

The cuddly mouse had turned into a fox. I was disgusted by his line of questioning. He was acting like a prosecutor than a judge whose job was to find the truth. His Foxship was trying to squeeze confession from a disorientated old woman.

Mrs Marsh replied calmly. 'I do not torment anyone, me Lord. I'm a good woman.'

'Who do you employ, then?'

'I employ nobody.'

'Who is it then?'

'How do I know?'

'The gaoler has said that the Devil visits you every night in your cell.'

'He visits me every night and brings me a jug of ale.'

'Did the Devil bid you to serve him?'

'No.'

'Did you make a covenant with him?'

'No.'

'Did you sign his book?'

Mrs Marsh was silent.

'Did you sign it with your blood?'

'No.'

'What evil spirits you are familiar with?'

'None.'

'What does the names Tig, Bik, Riv, Nim and Pif mean to you?'

'Nothing.'

'Which of these spirits you sent to bewitch Mr Donne to death?'

'None.'

'You are now in the hands of authority. Do you confess to your crimes and place yourself in God's merciful hands?'

'No, my Lord.'

'Who is your God?'

She placed her hand on the bailiff's Bible. 'The same as yours.'

'If you be guilty, pray God discovers you.'

I liked His Foxship's last comment but detested his bullying questioning. He asked the bailiff to take Mrs Marsh back to the dock, and then addressed the jury. 'Gentlemen of the Jury, this land is full of witches. I have hanged five or ten or twenty of them; there is no man here who can speak more of them than myself. You have heard what the witnesses have said about the prisoner. You have also heard what the prisoner can say for herself. Have an eye to your oath and to your duty, and do which God shall put in your minds to the discharge of your conscience, and mark well what is said.'

The clerk asked the bailiff to place his hand on the Bible. 'You shall well and truly keep every person sworn of this jury together in some private and convenient room without meat, drink, fire, candle or lodgings, and you shall not allow any person whatsoever to speak to them or any of them, neither shall you yourself speak to them until such time as they be agreed of their verdict, unless to ask them whether they have agreed of their verdict.'

Cosmo buried his head in the scroll. 'I have faith in Mr Johnson. The old lion is not going budge on his decision.' I was more worried about his health. 'Would he be allowed to take his medicines into the jury room?'

The sceptical chemist turns alchemy into chemistry, but fails to apply his scepticism to superstition.

Born in 1627 in English-occupied Ireland, Robert Boyle was the seventh son of the Earl of Cork. He was an extraordinarily bright child. At the age of fourteen he visited Florence to study the works of Galileo; Galileo died while Boyle was there. Galileo's works impressed him so profoundly that he decided to spend his life in science.

In 1661 he published his most famous work, *The Sceptical Chymist*, in which he rejected Aristotle's notion of four elements – earth, water, fire and air – and proposed that an element was a material substance and that it could be identified only by experiment. He presented the first scientific definition of an element when he said that elements were 'certain primitive and simple, or perfectly unmingled bodies which not being made of other bodies'. In other words, elements are one of the simplest components of matter, which could not be converted into anything simpler. He also said that elements were 'incapable of decomposition' – and added the prophetic – 'by any means with which we are now acquainted'.

In 1662 he made an efficient vacuum pump which he used to establish the relationship between the mass, volume and pressure of gases, known as Boyle's law: the volume of a given mass of a gas at constant temperature is inversely proportional to its pressure. He also used his pump to experiment on respiration and combustion and showed that air was necessary for life as well as for burning.

The famous English diarist Samuel Pepys records that King Charles – who himself dabbled in science and had his own private laboratory – 'mightily laughed' when told that scientists were 'spending time only in weighing air'. The cause of royal delight was Boyle's experiments on gases that led to his famous law.

Boyle, who was once introduced to Pepys as 'son of the Earl of Cork and father of modern chemistry', established chemistry as a science – an experimental science. He believed in 'setting up experiments and making observations' and not 'proclaiming any theories without having tested the relevant phenomena'.

In 1645 a group of men, including Boyle, met at Gresham College, London. The group continued to meet regularly for philosophical discussions and became known as the Invisible College. In 1660 the group won the support of King Charles and two years later it was incorporated by the royal charter into the Royal Society and its members became Fellows. The charter gave the society the right to publish under its own name, and in 1665 its journal *Philosophical Transactions* began publishing. The history of science since then is closely connected with the prestigious society and its journal.

'Elements make life; they can also end life,' Cosmo whispered. We're sitting on the floor leaning against the back wall of the courtroom.

'Are you composing a new *Pensées*?' I said.

The sugary, flirty, cheeky lass put her finger on her lips. 'Shhh!'

The bailiff had come into the courtroom; he was in a frantic state, shouting. 'Where is Dr Digby? Mr Johnson has taken ill.' The news crushed me, as if I was run over by a barley wagon. Cosmo face was surprisingly serene. He continued reading.

Why did demons torment science for so long? They may be out of science, but they're still worshipped in pseudoscience.

Boyle changed alchemy into chemistry, yet he was typical of men of his time. Like many of them he believed in alchemy, magic and witchcraft and was interested in establishing the reality of a supernatural realm. To seek evidence for a supernatural realm, he started dabbling in alchemy and even published an alchemical book. He also collaborated in witchcraft research with Joseph Glanvill, a controversial figure in the history of science. Boyle encouraged Glanville to think of witchcraft as a proper subject for science.

Glanvill was a member of the Royal Society and wrote many scientific works but is better known for a book with a funny title, *Sadducismus Triumphatus*, in which he defended witchcraft. He claimed that the stories of witchcraft 'contain nothing but what is consonant to right reason and sound philosophy'.

He believed that an important feature of the Society's work was 'the searching out of the true laws of matter and motion in order to securing the foundations of religion against all attempts of mechanical atheism.' Mechanical atheists, to him, were those scientists who were only interested in the physical character of the universe and encouraged a disbelief in the operation of any spiritual principles. God was not only essential to the creation of the universe, he said, but also to its continued existence. He proposed a scientific study of the supernatural world, which was not a well discovered region like America, and stood in the map of human science like unknown tracts.

Like astrology and alchemy, demonology, or belief in demons, is a pseudoscience, ideas and beliefs which masquerade as science, but have no or little relationship to scientific method. Demonology is incompatible with scientific insights which are gained through scientific methods. Why have demons lasted so long? The answer is superstition or irrationality of the human mind. But this is not the complete answer. Demons also contributed to an alliance between philosophy and religion.

Boyle started the process of separating science from pseudoscience. The process is now unstoppable. The real damage to the demons will come when future scientists discover the workings of nature. Fairies, witches, demons, astrology and alchemy would no longer be taken seriously by wise men – I hasten to add, Jane – and wise women.

Jane and Cosmo, my story of science has ended, but not the story of science; it'll continue as long as intelligent beings inhabit this world or any other world. Adio!

~~~

After half an hour the court was reconvened and the jurymen were back in the jury box. Salus was standing near the judge and whispering in his ear. The clerk stamped his white staff on the ground. His Foxship spoke. 'I'm afraid I have to announce sad news. Master Johnson has had acute pains in his chest. This court prays to God that he recovers fully.' He turned towards the jury box. 'You shall go back to the jury room and shall continue your deliberations without the benefit of the experience of Master Johnson.'

Cosmo slumped on the floor, tears streaming out of his eyes. I hugged him. I knew that the lyrical poetry of our dreams was about to be recast into the stark prose of reality. It took only fifteen minutes.

The jurymen were again in the jury box, the judge in his chair. The bailiff dragged Mrs Marsh to the bench. His Foxship held her hand and said to the jury. 'Look upon the prisoner you that be sworn, what say you, is she guilty or not guilty?'

Scarecrow stood up – my heart nearly stopped, if he was the foreman of the jury what hope was there for not-guilty verdict –

and said in a loud, bragging voice. 'Guilty on all counts, Your Lordship.' He waited for a few moments. 'May I say, Your Lordship, that the conscience of the jury is well satisfied and the witch is guilty and deserves death.'

The clerk stood up. 'Gaoler, lock the prisoner.'

His Foxship put a black cap on his head, shook his fur, focused his cunning eyes on his pray and jumped to kill the doomed victim. 'The prisoner shall be led back again to the place from where she came, and from there shall be taken to the place of execution, and there shall be hanged by the neck until she shall be dead.'

Someone shouted from the public gallery. 'Well done, Judge.' Applause that followed deafened my ears. My bright, cheerful young world had been sucked into a dull, dreary old black hole.

~~~

Friday. The execution day. Justice delayed is justice denied, but in my century the dispensing of justice was faster than the god of the storm. It took the grand jury one hour to commit an innocent woman to trial, the court six hours to find her guilty. After thirty-six hours roaring spectators, young and old, were waiting at the old castle for her to be hanged. I wished I could proclaim that justice, wise men's justice, should be delayed forever.

A carpenter had built a temporary scaffold over a ditch in a courtyard to hang the noose, a roper a noose that tightened slowly prolonging the excruciating pain of horrible death. A hangman was ready to force the condemned woman to stand on a platform, pull a handkerchief over her eyes, place the noose around her neck and kick away the platform.

Vicar was there to bless her soul. Sir Henry was there to give the signal for execution. Dr Digby was there to pronounce her dead when she had been brought down from the gallows after about an hour.

Only my Cosmo wasn't there. I had not seen him since the verdict. If you ask me how long eternity is, I would say it is definitely less than the past thirty-six hours of my life. Without Cosmo who would give the hangman the customary coin, who would provide a dignified closure to her life?

Broderick came running towards Sir Henry, shouting hoarsely. 'The witch is dead, my Lord, the witch has died in her cell.' The dandy, the vicar and the doctor rushed to her cell.

The girl's spirit was also in the cell reliving and relishing the moment when her friend took the solemn oath, 'I'll never let them hang you.' The cell was warm and bright, filled with cherubs with little golden swords in their hands to cast an eternal circle around a life as innocent as an unborn baby caught in a whirlpool of superstition. They all looked like newly born Cosmo to the little girl lost in a world where wise men had blanketed spiritual light of reason.

~~~

Months had passed and I was still hoping that Cosmo would come back; so was Mama who loved him like the son she never had. She had changed too; never mentioned Vicar Tayler, refused to go to the church with Salus. I refused to go to his rooms, the medical rooms that had turned medicine upside down into magic.

One day he came to my room with an empty bottle in his hand. 'Jane, did you ever see Cosmo taking medicine out of this bottle?'

When I didn't reply, he walked out of the room mumbling. 'It's alright if he gave it to the old woman, I hope he has not taken it himself. You see I found a bottle in her cell. Someone must have pushed it through the window.'

~~~

I still wake up in the middle of night. I see an old woman standing near my bed. I know she has woken up momentarily from the big sleep my Cosmo had blessed on her. 'You're a good lad, James,' she says lovingly and then dissolves into the darkness singing mockingly:

So listen girls to my song
The hangman's swung open his rope,
And on these gallows has been done
An end to Satan's hope.
Give the news from Hertford town
To all the world to spread,
An evil witch has gone down
Hanged by the neck, she is dead.

Closing the Circle

'That's my story, story of my days, Anne.' She was fast asleep. Had I been delivering my monologue, the mirror of my soul, without the requisite minimum audience for monologues? The signpost outside Ware that welcomed loons like me flashed before my eyes.

~~~

Next morning Anne came to my room with a scroll in her hand. It was tied with a blue string like Thales' scrolls.

'It's from Cosmo,' I said coolly as if I was expecting it.

There was surprise in Anne voice. 'Isn't he dead?'

'No. A saw him yesterday at the funeral.'

'Great news! Wasn't he Thales' spirit?'

'I thought you were sleeping.'

She laughed. 'A witch's words can hit you through dreams and darkness.'

'Cosmo didn't write the scrolls though he loved reading them.'

'Who had been writing them?'

'Thales' spirit?'

'I don't believe in such nutty things.'

'Would you believe if I said that I wrote them in the night without ever remembering in the morning that I had written them? I was possessed by a spirit. Even now I can't recall a thing.'

'It wasn't a supernatural spirit playing tricks on your mind. You know it, Auntie Jane. Your mind was obsessed about the idea that science would inspire people around you to follow the path of reason.'

'I failed.'

'No. Your life as a talented right-hand of a renowned astronomer has been inspiring. Now your Thales has made me walk, no, run on science's path.'

'Really! I wasn't lying when I said I couldn't read mirror writing.'

'It's not that difficult, my dear Auntie, I'll read Cosmo's scroll to you.'

The god of gravitation gives the laws of the universe, studies the nature of light, invents the refracting telescope and calculus, ...

No story of science is complete without Isaac Newton, the giant of science who said that if he had seen further it was by standing on the shoulders of giants. The old giants like Aristotle and Archimedes and the new ones like Kepler and Galileo, who died in 1642, the year of Newton's birth.

Say 'Newton' and the word 'gravitation' would spring to mind. Newton did his formative work on gravitation when he was twenty-three, in 1665–66 when the Great Plague had shut down Cambridge University, where a few months earlier he had completed his BA degree.

He published his findings in 1687 in his magnum opus *Philosophiae Natural Principia Mathematica* (Mathematical Principles of Natural Philosophy), in which he showed that one body attracts another body according to their mass and the distance which separates them. This universal force, gravitation, instantly affected everything in the universe, from particles of matter to apples to planets. The force exerted by the earth on an apple was the same force exerted by an apple on the earth. This force defies common sense, but it does keep the planets in orbit around the sun, causes apples to fall and holds your chair to the floor. Why do two objects attract each other? The great thinker couldn't find any explanation. 'I frame no hypotheses,' he said.

In *Principia* he also published his three laws of motion which changed the world, at least in the sense that every student of science has to learn them. The first law introduces the concept of inertia, the tendency of a body to resist change (which explains why you are suddenly pushed forward, if standing in a moving vehicle that suddenly stops). The second law explains the relationship between force and acceleration (a force is necessary to accelerate a body). The third law shows that forces always exist in pairs (for every action, there is an equal and opposite reaction).

In 1666 Newton tried a simple but ground-breaking experiment. He allowed a beam of sunlight to pass through a small round hole in a window shutter of a darkened room and placed a prism, which he had bought at a country fair, in the path of the beam. He was expecting a circle of white light on the opposite wall, and was quite surprised to see instead a band of seven rainbow colours. When he passed the spectrum through a second prism the seven colours were recombined into white light. He correctly concluded that sunlight is not homogenous and is made up of different colours. He continued his work on the nature of light and later invented the reflecting telescope (based on mirror) when he found the refracting telescope (based on lenses) would never give a good image.

He invented calculus as early as 1665, but he did not publish anything until 1687. Independent of Newton, the German mathematician and philosopher Gottfried Leibniz also arrived at the same method of dealing with variable quantities.

What was the key to his extraordinary creativity?' Chara wonders.

'Deep concentration and a fantastic dedication to work,' I replied.

Chara looked at me in admiration. 'Like you Thales.'

Newton's style of thinking was unique; his mind could easily abandon the familiar territory of 'common sense' and wander into new lands connecting a myriad of seemingly unconnected phenomena. He would sit down with a problem for hours unconscious that he had eaten his meals. When he was an old man and revered as a god, someone asked him the secret of his discoveries, he replied, 'Truth is the offspring of silence and unbroken meditation.'

His concentration was legendary so was his absent-mindedness. When at the age of twenty-seven he became Lucasian Professor he was expected to read a lecture on mathematics each week. He was so preoccupied with his work that he disregarded this obligation far more than he fulfilled. When he did lecture students were scarce. Sometimes he would read to a bare room.

... and searches for God in alchemy and theology. He was the first of the age of reason, but also the last of the sorcerers.

Newton established a mechanical universe which replaced the medieval universe. A Newton of future would replace the Newtonian universe with a better model. This process of discovery will continue until men and women of science have discovered the ultimate reality.

Like many scientists of his day Newton was fascinated by alchemy and spent much time and energy on his quest for the search of the philosopher's stone to turn base metals into gold. Not surprising, when you consider he was born in the superstitious 1640s, the decade in which, Matthew Hopkins, a man known as witch-finder general, had led the most chilling witch-hunt in English history which had resulted in hanging of more than hundred women.

Newton didn't believe in evil spirits and dismissed them as merely desires of the mind. He had private but intense interest in the study of theology and wrote over a million worthless words on mystical passages of the Bible. He saw divine fingerprints everywhere and believed that his work had demonstrated that the world must have been made by God. This supremely exquisite structure that is visible to us, he said, could come into being solely through the decision and under the dominion of a Supreme Being.

Newton's death in 1727 marked the beginning of a new era when science started sailing away towards the new world of reason, leaving the old island of theological dogma behind.

Demons are still living on the fringes of science worming into its exquisite fabric, says doubting Chara.

Chara glared glacially at me. 'You're wrong Thales, absolutely wrong. Even now there are men and women around who would argue with conviction that the earth is flat. Science and religion are still at war. Just listen to dumb debates like Creation vs Evolution and God vs science.'

'Smile Chara, smile; your name means happiness, and happiness abounds when you're closer to God. God is not a concept invented by Greek philosophers or other philosophers for that matter. We're born to believe in God. Almost every civilisation believes in some sort of supernatural power. Spirituality and morality are innate aspects of human nature.'

Chara scowls. 'You speak like a man of cloth.'

'No, as a scientist. Spirituality is a journey into our internal world to discover who we really are; science is an inquiry into the natural world around us, the seen and unseen, to discover where we are. Both seek one and the same truth, but in different ways. God is not irrelevant to science.'

'So, the moral of Thales' sermon from the ditch is that scientists should love God and their neighbours? If I pull you out of the ditch, again, you will see religion has always been anti-science and still is. Religion is not spirituality.'

'The essence of religious experience is spirituality, and the essence of spiritual experience is God.

Religion makes us stronger. It provides solace and guidance and helps heal wounds not only in our minds but also in our bodies. Most importantly, through prayers, rituals and orderliness it provides moral order that keeps society together.'

'May be you're right, Thales, but religion without reason is fundamentalism, which divides society.'

'Yes, I agree with you. Science transcends cultural and religious boundaries and thus encourages the oneness of humanity.'

'You mean both religion and science are good for us?'

'Yes Chara. No debate can prove otherwise because something intangible and inexplicable exists outside science's boundaries of matter, space and time. The cosmos is not self-caused, self-dependent, self-maintaining. To me, science is a religious cosmic experience because science also investigates phenomena outside human experience.'

'But don't let superstition corrupt your cosmic experience.' Chara yawns. 'It's time to go, Thales. Say adio to Jane and Cosmo and ask them to join us in a prayer for the victory of science over superstition in any future court of His Mouseship.'

I sobbed. 'Cosmo has lost his faith in science but not his faith in the power of prayer.'

'It's time you shared his faith. The spirits of Thales and Chara should also share two homes, the minds of a scientist and a priest.'

'What're you trying to prove?'

'Science and religion can exist in harmony. I'm going to the Abbey right now to bring Cosmo home. You know, Auntie Jane, in my century women can make things happen without casting a circle.'

~ ~ ~

Though the witch didn't pass on her powers to me because she was a witch with no magic and no wand, I'm her heiress, red-haired like her; and the long willow stick I once found to scare away a hedgehog who for a fleeting moment I thought was her is my magic wand.

With that wand I drew the circle around you, I'll now use it to close it ... from the east Amaymon, king of the east; from the north Zimimar, king of the north ... The circle has been closed. You may now leave.